Adrian Hofer

Spectroscopic studies in solid helium

Adrian Hofer

Spectroscopic studies in solid helium

Spectroscopic, time-resolved, and Stark effect studies of alkali atoms and exciplexes in solid 4He

Südwestdeutscher Verlag für Hochschulschriften

Impressum/Imprint (nur für Deutschland/ only for Germany)
Bibliografische Information der Deutschen Nationalbibliothek: Die Deutsche Nationalbibliothek
verzeichnet diese Publikation in der Deutschen Nationalbibliografie; detaillierte bibliografische
Daten sind im Internet über http://dnb.d-nb.de abrufbar.
Alle in diesem Buch genannten Marken und Produktnamen unterliegen warenzeichen-, marken-
oder patentrechtlichem Schutz bzw. sind Warenzeichen oder eingetragene Warenzeichen der
jeweiligen Inhaber. Die Wiedergabe von Marken, Produktnamen, Gebrauchsnamen,
Handelsnamen, Warenbezeichnungen u.s.w. in diesem Werk berechtigt auch ohne besondere
Kennzeichnung nicht zu der Annahme, dass solche Namen im Sinne der Warenzeichen- und
Markenschutzgesetzgebung als frei zu betrachten wären und daher von jedermann benutzt
werden dürften.

Verlag: Südwestdeutscher Verlag für Hochschulschriften Aktiengesellschaft & Co. KG
Dudweiler Landstr. 99, 66123 Saarbrücken, Deutschland
Telefon +49 681 37 20 271-1, Telefax +49 681 37 20 271-0, Email: info@svh-verlag.de
Zugl.: Fribourg, Schweiz, Universität Freiburg, Departement für Physik, Dissertation 1571, 2007

Herstellung in Deutschland:
Schaltungsdienst Lange o.H.G., Berlin
Books on Demand GmbH, Norderstedt
Reha GmbH, Saarbrücken
Amazon Distribution GmbH, Leipzig
ISBN: 978-3-8381-0952-7

Imprint (only for USA, GB)
Bibliographic information published by the Deutsche Nationalbibliothek: The Deutsche
Nationalbibliothek lists this publication in the Deutsche Nationalbibliografie; detailed
bibliographic data are available in the Internet at http://dnb.d-nb.de.
Any brand names and product names mentioned in this book are subject to trademark, brand
or patent protection and are trademarks or registered trademarks of their respective holders.
The use of brand names, product names, common names, trade names, product descriptions
etc. even without a particular marking in this works is in no way to be construed to mean that
such names may be regarded as unrestricted in respect of trademark and brand protection
legislation and could thus be used by anyone.

Publisher:
Südwestdeutscher Verlag für Hochschulschriften Aktiengesellschaft & Co. KG
Dudweiler Landstr. 99, 66123 Saarbrücken, Germany
Phone +49 681 37 20 271-1, Fax +49 681 37 20 271-0, Email: info@svh-verlag.de

Copyright © 2009 by the author and Südwestdeutscher Verlag für Hochschulschriften
Aktiengesellschaft & Co. KG and licensors
All rights reserved. Saarbrücken 2009

Printed in the U.S.A.
Printed in the U.K. by (see last page)
ISBN: 978-3-8381-0952-7

Abstract

This thesis work was carried out in the atomic physics group of the Physics Department of the University of Fribourg. The thesis is divided into two main parts, and contains in total six original research papers. The first part deals with purely optical studies of Cs and Rb atoms and their exciplex molecules and clusters implanted in solid ^{4}He crystals. The second part is an experimental and theoretical study of the Stark effect in the ground state of Cs atoms implanted in the bcc phase of solid ^{4}He.

We have used the well known technique of laser ablation implantation of defect atoms into solid ^{4}He. During the ablation process not only atoms but also charged particles, molecules, and metallic clusters are ablated and become trapped in the He crystal. The doped part of the crystal, which we call an "iceberg", has a cylindrical shape and an extension of a few centimeters in the vertical direction. An interesting phenomenon occurs during the melting of the crystal in that the iceberg stays solid even when the surrounding He is already liquid. We have studied the absorption spectrum of the iceberg and performed interferometric measurements of its structure to learn about its index of refraction. From those measurements we infer that the He density in the iceberg lies between the densities of liquid and bcc solid He. We speculate that the iceberg is held together by ions and electrons. The ions form so-called snowballs, whereas the electrons form cavities (electron bubbles) due to the Pauli repulsion; it seems to be energetically favorable that the ions and electrons do not recombine and build an amorphous or crystalline ionic structure. (Paper I)

The spectral shift of atomic absorption and emission lines due to the interaction of the alkalis with the He matrix has been studied extensively in the past and many features could be explained by the so-called bubble model. However that model could not explain the sudden jumps of the absorption and emission wavelenghts at the liquid-solid phase transition. In this work we have extended the model by including an additional force in the solid phase to account for the elastic properties of the crystal. The extended model also accounts for the spectral shift due to an electromagnetic cavity, an effect not considered in earlier models. The cavity effect is a manifestation of the interaction of the atomic dipole with its own mirror images in the surrounding dielectric. Extensive measurements of the pressure shift of atomic absorption and emission lines are compared with our model calculations. (Paper II)

The reflected field of the atomic dipole (cavity effect) leads not only to a frequency shift of the emitted light, but also to changes in the radiative lifetime of the excited state. In this work we carried out lifetime measurements of the Cs $6P_{1/2}$ state in bcc and hcp solid ^{4}He and studied their pressure dependence. The values in bcc coincide with previous measurements performed by another group in the liquid phase. The lifetimes are compared to our theoretical predictions from the extended bubble model. The smaller lifetime in liquid and bcc solid ^{4}He compared to the free atomic value is well reproduced by the model only if the cavity effect is taken into account. The pressure independence of the lifetime in superfluid and bcc solid ^{4}He can be explained by the model and is due to the compensation of two effects: the reduction of the transition dipole matrix elements with increasing helium pressure and the increase of the frequency of the emitted light. An additional nonradiative decay channel opens in the hcp phase and leads to a sudden jump of the lifetime at the bcc-hcp phase transition. The additional decay channel had been studied before in our group and consists in the formation of Cs*He$_n$ ($n = 2$, $n = 6 - 7$) exciplexes. (Paper III)

Exciplexes, i.e., bound states between an excited alkali atom and one or several He atom(s), are interesting objects in themselves. In this thesis the previous Cs exciplex studies were extended to the Rb system. The smaller fine-structure splitting of Rb compared to Cs changes the exciplex formation probability. The pressure dependent quenching of the Rb D_1 emission in superfluid He was explained before as being due to exciplex formation. With Cs, no exciplex formation could be observed in superfluid and in bcc solid He following D_1 excitation (excitation to the $6P_{1/2}$ state), confirmed by our lifetime measurements of the Cs $6P_{1/2}$ state. Our experiments on Rb atoms in bcc and hcp solid ^{4}He have shown that the strongest decay channel of Rb atoms excited to the $5P_{1/2}$ and $5P_{3/2}$ states is the formation of a Rb*He$_6$ exciplex. Weak emission from the two linear Rb*He$_1$ and Rb*He$_2$ exciplexes was also observed. The theoretical model developed for the Cs exciplexes has been applied with success to the Rb-exciplex system and has allowed us to identify all of the observed emission lines. During these experiments we have also observed for the first time a faint emission from the Rb D_1 and D_2 atomic lines in solid ^{4}He. (Paper IV)

Part two of the thesis studies the effect of a static electric field on the properties of the Cs ground state. The quadratic Stark effect leads to a global shift, quadratic in the applied electric field strength, of the magnetic sublevels in the Cs ground state and can be parameterized by a scalar polarizability α_0. There is however a tiny contribution $\alpha_2^{(3)}$ (tensor polarizability) to the scalar polarizability ($\alpha_2^{(3)} \approx 10^{-7} \alpha_0$) which lifts the Zeeman degeneracy of the hyperfine sublevels. In this work we report experimental details of the Stark effect measurements on Cs atoms implanted in the bcc phase of solid ^{4}He. Optically detected magnetic resonance (ODMR) was used to detect tiny shifts of the magnetic sublevels in the Cs ground state due to the applied static electric field. The experimental value for the tensor polarizability $\alpha_2^{(3)}$ of the Cs ground state in bcc solid ^{4}He differs from the free atomic value by 10%. (Paper V)

The extended bubble model was used to calculate the wavefunctions and energy levels of the Cs atom in the bubble. These quantities were then used to evaluate numerically the perturbation expansion to calculate the influence of the He matrix on the tensor polarizability. We show that the theoretical value is in good agreement with the experimental data. An extensive theoretical paper treats also the free atom for which the Schrödinger equation with a scaled Thomas-Fermi model potential was solved to calculate the wavefunctions of the free Cs atom up to principal quantum numbers n=200. These wavefunctions were used, as for the case in the bubble, to calculate the tensor polarizability. We showed by explicit calculation of continuum wavefunctions, that their influence can be neglected. We conclude that theoretical and experimental values of the tensor polarizability $\alpha_2^{(3)}$ are in good agreement for the free Cs atom and also for the Cs atom in bcc solid ^{4}He. The third order perturbation theory used to calculate $\alpha_2^{(3)}$ (developed by my former colleague S. Ulzega and myself and presented in the Ph. D. thesis of S. Ulzega and in this work) could bridge the 40-year old gap between theory and experiment for the value of $\alpha_2^{(3)}$ in the free atom. (Paper VI)

Zusammenfassung

Die vorliegende Arbeit wurde in der Atomphysik Gruppe des Departements für Physik der Universität Freiburg verfasst. Die Arbeit ist in zwei Hauptteile gegliedert, welche insgesamt 6 wissenschaftliche Publikationen enthalten. Im ersten Teil werden optische Untersuchungen an Cs und Rb Atomen sowie deren Exciplex-Molekülen und Clustern in festen ^{4}He Matrizen präsentiert. Der zweite Teil stellt experimentelle und theoretische Untersuchungen des Stark-Effektes im Grundzustand von Cs Atomen in der bcc Phase von festem ^{4}He vor.

Wir haben die bekannte Technik der Laserablation benutzt um Fremdatome in einen Helium-Kristall einzupflanzen. Während der Ablation werden nicht nur Atome, sondern auch geladene Teilchen, Moleküle und metallische Cluster abgelöst und anschliessend im Heliumkristall gefangen. Der Teil des Kristall, der diese Verunreinigungen enthält, nennen wir „Eisberg" auf Grund seiner Form. Er ist zylindrisch und ist ein paar Zentimeter in vertikaler Richtung ausgedehnt. Eine interessante Eigenschaft dieses Eisberges ist, dass er fest bleibt wenn das ihn umgebende Helium bereits flüssig ist. Wir haben Absorptionsspektren und interferometrische Untersuchungen an diesem Eisberg durchgeführt. Aus diesen Messungen können wir schliessen, dass die Heliumdichte im Eisberg zwischen der Dichte von flüssigem und festem Helium in der bcc Phase liegt. Wir spekulieren, dass der Eisberg durch Ionen und Elektronen zusammengehalten wird. Die Ionen bilden so genannte Schneebälle, während die Elektronen, auf Grund der Pauli-Abstossung Blasen (Elektronenblasen) bilden. Energetisch scheint es vorteilhaft zu sein, wenn die Ionen und Elektronen nicht rekombinieren und somit eine amorphe (oder kristalline) ionische Struktur bilden. (Publikation I)

Die spektrale Verschiebung von atomaren Absorptions- und Emissionslinien auf Grund der Wechselwirkung der Alkali-Atome mit der Helium Matrix ist in der Vergangenheit ausführlich untersucht worden und viele Beobachtungen konnten mit Hilfe des so genannten Blasenmodells erklärt werden. Dieses Modell kann jedoch die Sprünge der Absorptions- und Emissionswellenlänge an der Phasengrenze von flüssigem zu festem Helium nicht erklären. In dieser Arbeit wurde das Modell erweitert, in dem eine zusätzliche Kraft für die feste Phase eingeführt wurde, die die elastischen Eigenschaften des Kristalls berücksichtigt. Das erweiterte Modell beinhaltet auch eine zusätzliche spektrale Verschiebung der Linien, ein Effekt, der durch die Blase, in der sich die Alkali-Atome befinden, entsteht (Blaseneffekt). Hierbei wechselwirken die atomaren Dipole der angeregten Atome mit ihren eigenen Spiegelbildern, welche durch die dielektrischen Blasen erzeugt werden. Ausführliche Messungen der Druckverschiebung von atomaren Absorptions- und Emissionslinien wurden erfolgreich mit unseren Modellrechnungen verglichen. (Publikation II)

Das reflektierte elektromagnetische Feld des atomaren Dipols bewirkt nicht nur eine Frequenzverschiebung (Blaseneffekt), sondern führt auch zu einer veränderten radiativen Lebensdauer des angeregten Zustandes der Atome. Wir haben während dieser Arbeit Lebensdauer-Messungen des Cs $6P_{1/2}$ Zustandes in der bcc und hcp Phase von festem Helium durchgeführt und haben die Druckabhängigkeit dieser Lebensdauer untersucht. Die Werte, die in der bcc Phase gemessen wurden, stimmen mit früheren Messungen in der flüssigen Phase von einer anderen Forschungsgruppe überein. Die Lebensdauern werden mit den theoretischen Vorhersagen des erweiterten Blasenmodells verglichen. Die kürzere Lebensdauer in flüssigem und festem (bcc) Helium werden durch die Modellrechnungen gut wiedergeben, wenn man den Blaseneffekt berücksichtigt. Die druckunabhängige Lebensdauer in der flüssigen und der bcc Phase kann mit dem Modell erklärt werden und beruht auf der Kompensation zweier Effekte: das Dipol Matrixelement nimmt mit zunehmendem Druck ab, währen die Frequenz des abgestrahlten Lichtes zunimmt. In der hcp Phase des festen Heliums existiert ein zusätzlicher nichtstrahlender Zerfallskanal, der zu einem Sprung der Lebensdauer an der bcc-hcp Phasengrenze führt. Dieser zusätzliche Zerfallskanal wurde bereits früher in unserer Forschungsgruppe untersucht und besteht in der Bildung von Cs*He$_n$ (n=2, $n = 6 - 7$) Exciplex-Molekülen. (Publikation III)

Exciplexe, d.h. gebundene Zustände zwischen einem angeregten Alkali-Atom und einem oder mehreren Helium Atomen, sind interessante Objekte. In dieser Arbeit wurden die früheren Untersuchungen der Cs-Exciplexe auf das Rubidium System ausgedehnt. Rb hat im Vergleich mit Cs eine kleinere Feinstrukturkonstante, weshalb die Wahrscheinlichkeit zur Exciplexbildung verändert wird. Die druckabhängige Dämpfung der Rb D_1 Emission in superfluidem Helium wurde schon früher durch die Bildung von Exciplexen erklärt. Bei Cs hingegen konnte keine Exciplexbildung in superfluidem und festem (bcc) Helium nachgewiesen werden, bei Anregung der Atome auf der D_1 Linie (Anregung des $6P_{1/2}$ Zustandes). Dies wurde auch durch unsere Lebensdauermessungen des $6P_{1/2}$ Zustandes bestätigt. Unsere Experimente mit Rb Atomen, die in der bcc und der hcp Phase von ^{4}He eingebunden sind, haben gezeigt, dass der stärkste Zerfallskanal der angeregten Rb Atome (Anregung des $5P_{1/2}$ oder $5P_{3/2}$ Zustandes) die Bildung von Rb*He$_6$ Exciplexen ist. Ein schwaches Fluoreszenzsignal wurde auch von den beiden linearen Exciplexen Rb*He$_1$ und Rb*He$_2$ beobachtet. Das theoretische Modell, dass für die Cs Exciplexe entwickelt wurde, wurde erfolgreich auf das Rb-Exciplex System angwandt und hat uns erlaubt, alle beobachteten Emissionslinien zu identifizieren. Während dieser Messungen haben wir auch zum ersten Mal eine schwache Emission der atomaren Rb D_1 und D_2 Linien in festem ^{4}He beobachtet. (Publikation IV)

Der zweite Hauptteil dieser Doktorarbeit untersucht den Einfluss eines statischen elektrischen Feldes auf die Eigenschaften des Cs Grundzustands. Der quadratische Stark-Effekt (quadratisch in der angelegten elektrischen Feldstärke) führt zu einer globalen Verschiebung der magnetischen Unterzustände im Cs Grundzustand. Dieser Effekt kann durch eine skalare Polarisierbarkeit α_0 parametrisiert werden. Es existiert aber ein winziger Beitrag $\alpha_2^{(3)}$ (genannt Tensorpolarisierbarkeit) zur skalaren Polarisierbarkeit ($\alpha_2^{(3)} \approx 10^{-7}\alpha_0$), der die Zeeman Entartung der Hyperfein-Unterzustände aufhebt. Wir präsentieren in dieser Arbeit experimentelle Details der Stark-Effekt Messungen, die an Cs Atomen in der bcc Phase von festem ^{4}He durchgeführt wurden. Wir haben optisch detektierte Magnetresonanz benutzt, um diese kleinen Verschiebungen der magnetischen Unterzustände, bewirkt durch das angelegte statische elektrische Feld im Cs Grundzustand zu messen. Der experimentelle Wert der Tensorpolarisierbarkeit $\alpha_2^{(3)}$ des Grundzustandes für Cs Atome in bcc festem ^{4}He weicht um etwa 10% vom Wert im freien Atom ab. (Publikation V)

Das erweiterte Blasenmodell wurde benutzt um Wellenfunktionen und Energieniveaus der Cs Atome in der atomaren Blase zu berechnen. Diese Grössen wurden für die numerische Auswertung der Störungsrechnung in dritter Ordnung benutzt, um den Einfluss der He Matrix auf die Tensorpolarisierbarkeit zu berechnen. Wir zeigen, dass der theoretische Wert gute Übereinstimmung mit dem experimentellen Wert zeigt. Die ausführliche theoretische Publikation behandelt auch das freie Atom, für welches ein skaliertes Thomas-Fermi Potential benutzt wurde, um die Wellenfunktionen des freien Cs Atoms bis zur Hauptquantenzahl n=200 zu berechnen. Diese Wellenfunktionen wurden, wie für die Rechnung in der Blase, benutzt, um die Tensorpolarisierbarkeit zu berechnen. Wir haben durch explizite Berechnung von Kontinuumswellenfunktionen gezeigt, dass deren Einfluss vernachlässigbar klein ist. Wir schliessen mit der Aussage, dass theoretische und experimentelle Werte der Tensorpolarisierbarkeit $\alpha_2^{(3)}$ für das freie Cs Atom wie auch für das Cs Atom im bcc festen ^{4}He in guter Übereinstimmung sind. Die Störungsrechnung dritter Ordnung, die zur Berechnung von $\alpha_2^{(3)}$ benutzt wurde (entwickelt von meinem führen Kollegen S. Ulzega und mir selbst und vorgestellt in der Doktorarbeit von S. Ulzega und in dieser Arbeit), konnte somit das Rätsel um die 40 Jahre bestehende Diskrepanz zwischen theoretischem und experimentellem Wert für $\alpha_2^{(3)}$ im freien Atom erfolgreich lösen. (Publikation VI)

Contents

Preface	1
I Introduction	**3**
1 General introduction to the experiments	**5**
1.1 Historical overview	5
1.2 Properties of liquid and solid ^{4}He	6
1.2.1 Phase diagram and structure of the two solid phases	6
1.2.2 He-He interactions	7
1.2.3 Molar volume of solid ^{4}He	8
1.2.4 Optical properties	10
1.3 Experimental setup	10
1.3.1 The bath cryostat	10
1.3.2 The pressure cell	11
1.3.3 Implantation process	12
II Optical spectroscopy of atoms and exciplexes in solid ^{4}He	**15**
2 Introduction to spectroscopic and time-resolved measurements in solid ^{4}He	**17**
2.1 Optical spectroscopy	17
2.1.1 Atomic bubble and optical absorption-emission cycle	18
2.2 Alkali dimers in solid He	18
3 Paper I: **Impurity-stabilized solid ^{4}He below the solidification pressure of pure helium.**	**23**
3.5 Methods	29
4 Paper II: **$6S_{1/2}$-$6P_{1/2}$ transition of Cs atoms in cubic and hexagonal solid ^{4}He**	**31**
4.1 Introduction	34
4.2 Theoretical model	34
4.2.1 Spherical bubble model	34
4.2.2 Energy of the free Cs atom	35
4.2.3 Cs-He interaction	36
4.2.4 Integration over the bubble and the bubble energy	36
4.2.5 Hyperfine structure of Cs in solid He	38
4.2.6 Fine structure of Cs in solid He	38
4.2.7 Lifetime of the $6P_{1/2}$ state	39

	4.2.8	The cavity effect	39
	4.2.9	Line shape of absorption and emission lines	40
4.3	Experiment		41
4.4	Discussion		42
	4.4.1	Extension of the bubble model	42
	4.4.2	Summary	43

5 Paper III:
Lifetime of the Cs $6P_{1/2}$ state in bcc and hcp solid ^{4}He   45

5.1	Introduction		48
5.2	Theory		48
5.3	Experiment		49
	5.3.1	Experimental setup	49
	5.3.2	Life time measurements	50
5.4	Discussion		51
	5.4.1	Lifetimes in the bcc phase	51
	5.4.2	Lifetimes in the hcp phase	51
	5.4.3	Analysis	52
	5.4.4	Summary	53

6 Paper IV:
Rb*He$_n$ exciplexes in solid ^{4}He   55

6.1	Introduction		58
6.2	Theory		59
6.3	Experimental results		60
	6.3.1	Experimental setup	60
	6.3.2	Atomic Bubbles	60
	6.3.3	Emission spectra following D_1 excitation	61
	6.3.4	Emission spectra following D_2 excitation	61
	6.3.5	Atomic and exciplex excitation spectra	62
6.4	Discussion		63
	6.4.1	Atomic lines	63
	6.4.2	Apple-shaped Rb(B$^2\Pi_{3/2}$)He$_{1,2}$ exciplexes	64
	6.4.3	Diatomic bubble	65
	6.4.4	Dumbbell-shaped Rb(A$^2\Pi_{1/2}$)He$_{n>2}$ exciplexes	65
	6.4.5	Formation of dumbbell-shaped Rb(A$^2\Pi_{1/2}$)He$_{n>2}$ exciplexes	66
	6.4.6	Summary and conclusion	67

III Magneto-optical experiments in solid ^{4}He   69

7 Introduction to magneto-optical experiments in solid ^{4}He   71

7.1	Magnetic resonance experiments		71
	7.1.1	Optically detected magnetic resonance (ODMR)	72
7.2	Principle of EDM and Stark effect measurements		72
	7.2.1	Stark effect	73
	7.2.2	General concept of an EDM measurement	73
	7.2.3	EDM measurement in solid He	75

8 Paper V:
Measurement of the forbidden electric tensor polarizability of Cs atoms trapped in solid ^{4}He   **79**

8.1 Introduction 82
8.2 Theory 83
8.3 Experimental methods 83
     8.3.1 Helium matrix isolation spectroscopy 83
     8.3.2 The sample cell 83
     8.3.3 The magnetic resonance technique 85
8.4 Measurements 85
     8.4.1 The tensor polarizability in the M_z geometry 85
     8.4.2 The relative sign of $\alpha_2(F=4)$ and $\alpha_2(F=3)$ 86
     8.4.3 Line drifts 86
     8.4.4 The tensor polarizability in the M_x geometry 87
8.5 Analysis of the M_x data 88
8.6 Comparison with theory 90
8.7 Summary 90

9 Paper VI:
Calculation of the forbidden electric tensor polarizabilities of free Cs atoms and of Cs atoms trapped in a solid ^{4}He matrix   **93**

9.1 Introduction 96
9.2 Theory 96
     9.2.1 Second order perturbation theory 97
     9.2.2 Third order perturbation theory 97
9.3 The third order polarizability of the free cesium atom 98
     9.3.1 Earlier calculation revisited 98
     9.3.2 Inclusion of off-diagonal hf matrix elements 99
     9.3.3 Electric dipole matrix elements 100
     9.3.4 Wavefunctions of the free cesium atom 100
     9.3.5 Terms with diagonal hf matrix elements 102
     9.3.6 Terms with off-diagonal hf matrix elements 103
     9.3.7 Contribution of continuum states 103
     9.3.8 Numerical evaluation of the third order tensor polarizability for the free Cs atom . 103
9.4 The third order polarizability of cesium in solid helium 104
     9.4.1 Experiment 104
     9.4.2 Wavefunction of Cs in solid ^{4}He 104
     9.4.3 Numerical evaluation of the third order tensor polarizability of Cs in solid He ... 105
9.5 Summary 107

Summary and Outlook   **109**

Danksagung   **111**

Curriculum vitae   **113**

Preface

The research carried out during this thesis was performed in the atomic physics group FRAP (Fribourg group for atomic physics) headed by Prof. Dr. A. Weis. It is focused on the optical and magneto-optical spectroscopy on ^{4}He crystals doped with Rb and Cs. The thesis contains a general introduction to the experiments with a short historical overview. It is a collection of 6 papers (five published, and one submitted) thematically grouped in two parts, which each contains a short introductory text.

The first part (four papers) deals with optical spectroscopic and time-resolved studies of Cs and Rb atoms as well as Rb*He$_n$ exciplexes implanted in a solid ^{4}He matrix.

The second part (two papers) presents experimental and theoretical studies, respectively, of the Stark effect in the ground state of Cs implanted in the bcc phase of solid ^{4}He.

Each paper can be understood as an independent text, containing an introduction, the main text, a summary and the relevant references. The work was done in collaboration with former Ph. D. students and post-docs, all of which have contributed to the research work. For this reason I briefly outline before each paper my own contributions to the work therein.

The articles included in this thesis are:

Part II

Chapter 3
Paper I: P. Moroshkin, A. Hofer, S. Ulzega and A. Weis. *Impurity-stabilized solid ^{4}He below the solidification pressure of pure helium.*, Nature Physics **3**, 786 (2007).

Chapter 4
Paper II: A. Hofer, P. Moroshkin, S. Ulzega, D. Nettels, R. Müller-Siebert and A. Weis. *$6S_{1/2}$-$6P_{1/2}$ transition of Cs atoms in cubic and hexagonal solid ^{4}He*, Phys. Rev. A **76**, 022502 (2007).

Chapter 5
Paper III: A. Hofer, P. Morohskin, S. Ulzega and A. Weis. *Lifetime of the Cs $6P_{1/2}$ state in bcc and hcp solid ^{4}He*, Europhys. J. D (2007), DOI: 10.1140/epjd/e2007-00275-5.

Chapter 6
Paper IV: A. Hofer, P. Moroshkin, D. Nettels, S. Ulzega and A. Weis. *Rb*He$_n$ exciplexes in solid ^{4}He*, Phys. Rev. A **74**, 032509 (2006).

Part III

Chapter 8
Paper V: S. Ulzega, A. Hofer, P. Moroshkin, R. Müller-Siebert, D. Nettels and A. Weis. *Measurement of the forbidden electric tensor polarizability of Cs atoms trapped in solid ^{4}He*, Phys. Rev. A **75**, 042505 (2007).

Chapter 9
Paper VI: A. Hofer, P. Moroshkin, S. Ulzega and A. Weis. *Calculation of the forbidden electric tensor polarizabilities of free Cs atoms and of Cs atoms trapped in a solid ^{4}He matrix*, submitted to Phys. Rev. A.

Part I

Introduction

Chapter 1

General introduction to the experiments

1.1 Historical overview

The field of classical matrix isolation spectroscopy of atoms in heavy rare gas matrices is about 40 years old. In these experiments the rare gas atoms and the atoms under investigation are condensed on a finger cooled to cryogenic temperatures. This method fails if one uses solid ^{4}He as the host matrix since He solidifies only under pressure even at the absolute zero of temperature.

In 1991 Kanorsky and Weis proposed to use paramagnetic atoms implanted in superfluid ^{4}He for the search of a permanent electric dipole moment of the electron (eEDM) [1, 2]. The search for an EDM of a fundamental particle like the electron or the neutron is today a very promising way to look for physics beyond the Standard Model of elementary particles. The EDM of a heavy paramagnetic atom can be induced by the EDM of an electron. The EDM of the electron is enhanced in paramagnetic atoms with respect to the free electron. The reason for proposing He as a host matrix is its diamagnetic character, expected to give very long spin relaxation times of the implanted atoms and therefore narrow magnetic resonance lines. Moreover superfluid He has a large electric breakdown voltage ($\geq$ 100kV/cm). It was suggested that these two properties would make superfluid He an ideal environment for a sensitive atomic EDM experiment.

In the following years Weis and Kanorsky succeeded at the Max-Planck-Institut für Quantenoptik (MPQ) to do first optical studies of Ba, Au and Cu atoms in superfluid He [3]. By means of laser ablation from a solid target, atomic number densities of $10^8 - 10^9$ cm^{-3} of immersed atoms were reached. The pressure shift of excitation and emission lines of Ba in superfluid He was quantitatively explained by the so-called spherical bubble model [4]. Due to the Pauli repulsion the implanted neutral atoms repel the He atoms and form a cavity (the atomic bubble).

In parallel to these experiments at MPQ the group led by T. Yabuzaki performed optical and magnetic resonance (MR) experiments on Cs in superfluid He [5]. They measured relatively broad MR lines (10^5 Hz) probably due to the limited time of observation (much less than one second) determined by the atoms diffusing out of the region of interest.

First implantations of Ba and Cs atoms into solid He by means of laser ablation were done in 1993 at MPQ [6]. Solid He is very well suited for high precision spin physics. The implanted atoms diffuse only very slowly out of the investigation region (few hours) in contrast to superfluid He (less than one second). Atoms in solid He are mainly lost due to dimer and cluster formation. A constant number of foreign atoms can be obtained by applying pulses from the same Nd:YAG laser used for the ablation process, thereby destroying the clusters. Longitudinal spin relaxation times of 1 second for Cs in solid He and magnetic resonance linewidths of 2 kHz (two orders of magnitude smaller than in superfluid He) were measured [7]. The lines were broadened by residual magnetic fields and later, linewidths of only 20 Hz were obtained [8] after reduction of magnetic field inhomogeneities. The studies on solid He were continued at the Institute for Applied Physics of the University of Bonn and, since 2000, at the Physics Department of the University of Fribourg, Switzerland.

Doped solid He is such a complex and unique system that it has led to many interesting discoveries on the way to the far-reaching goal of an EDM experiment. A review of the research of the team of Prof. A. Weis can be found in [9]. The idea of measuring an EDM of the electron with this system had to be abandoned recently for different reasons, that will be addressed in chapter 7. The research field is closely related to dopant spectroscopy in other quantum solids and fluids, such as solid hydrogen or helium nanodroplets.

1.2 Properties of liquid and solid ^{4}He

1.2.1 Phase diagram and structure of the two solid phases

Helium is an element with unique properties. It is the only element which stays liquid down to the absolute zero of temperature under its saturated vapor pressure. In 1908 the first liquid helium was produced by Heike Kamerlingh-Onnes [10]. A little later he discovered that liquid He can cover vertical surfaces (called the Onnes-effect). In 1926 Willem Hendrik Keesom found a method to solidify ^{4}He by applying pressure [11]. The superfluid phase (HeII) and the hcp (hexagonal close packed) structure of solid He were discovered in 1938 [12, 13, 14]. In the superfluid phase the viscosity tends to zero. In 1953 the solid fcc (face centered cubic) phase above 1000 bar was discovered [15]. Only in 1962 the small island of bcc (body centered cubic) phase at temperatures of 1.6 K and pressures of 28 bar was found [16, 17].

Figure 1.1 shows the phase diagram of ^{4}He. The normal liquid phase (HeI) starts at 4.21 K. Below the critical temperature ($T_c = 2.177$ K) He becomes superfluid (HeII). The λ line shown in Fig. 1.1 separates the normal from the superfluid phase. The solid phase can only be reached by applying pressures in excess of 25.3 bars.

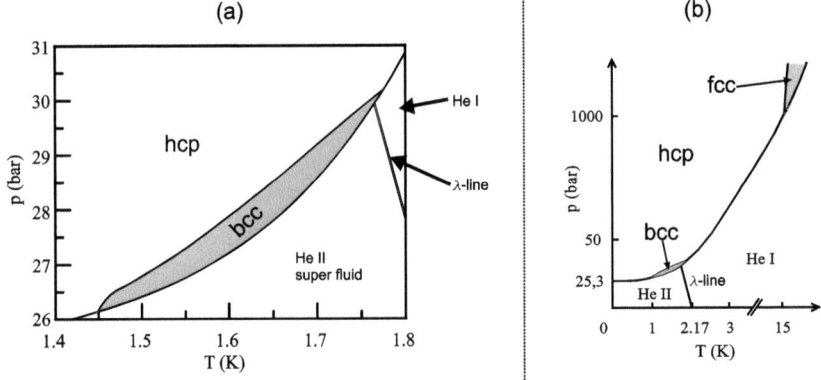

FIGURE 1.1: *Phase diagram of ^{4}He.*

Experiments in this work were carried out in the cubic bcc and the uniaxial hcp phase of ^{4}He. Figure 1.2 shows the primitive cells of these two solid phases. One sees that the hcp phase has an orientation and two different lattice constants.

The superfluid phase of He is a macroscopic manifestation of a quantum phenomenon. Helium atoms are bosons and can therefore undergo a Bose-Einstein condensation, where all particles condense into the same lowest energy state so that the macroscopic sample can be described by one single wavefunction. Such a phase transition occurs when the thermal de Broglie wavelength

$$\lambda_{dB} = \sqrt{\frac{2\pi\hbar^2}{m_{He}\,kT}} \qquad (1.1)$$

becomes comparable to the interatomic distance (~ 3 Å) in the liquid. At 1.5 K the thermal de Broglie wavelength of He is around 7 Å and the atoms are thus strongly delocalized.

1.2 Properties of liquid and solid ^{4}He

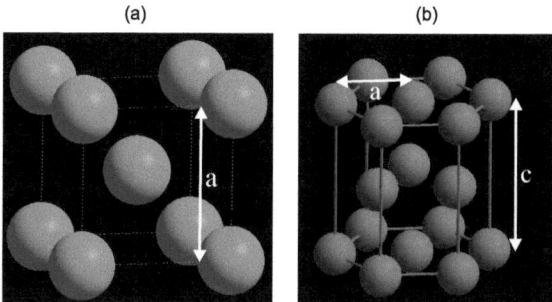

FIGURE 1.2: *Primitive cells of solid ^{4}He. (a) In the bcc phase with a lattice constant a=4.12 Å at T=1.62 K and p=27.9 bar and (b) in the hcp phase with lattice constants a=3.67Å and c=6.01 Å at T=1 K and p=27.15 bar [18, 19].*

A	$1.869244 \cdot 10^5$
α	10.5717543
C6	1.35186623
C8	0.41495143
C10	0.17151143
β	-2.07758779
D	1.438
ϵ (J)	$15.1265 \cdot 10^{-23}$
r_m (Å)	2.6413813

TABLE 1.1: *Parameters used for the analytical representation of the He-He interaction.*

Solid He is a so-called quantum crystal because the zero point energy of the He atoms is comparable to their potential energy (details see Sect. 1.2.2). The strong overlap of the wavefunctions of individual He atoms gives the crystal a macroscopic quantum nature.

1.2.2 He-He interactions

The He-He interaction at short internuclear distances is dominated by a repulsive Pauli interaction due to the closed 1^2s shell of the He atoms. At larger internuclear distances the attractive van der Waals interaction leads to a very shallow potential well. The He-He interaction used in this work is taken from [20]. The potential is an analytical representation of an *ab initio* potential using the Hartree-Fock-dispersion form. It is the most accurate potential for the He-He interaction today whose analytical form is

$$V_{He-He} = \epsilon \left\{ A\, e^{-\alpha \frac{r}{r_m} + \beta \left(\frac{r}{r_m}\right)^2} - \left[C6 \left(\frac{r_m}{r}\right)^6 + C8 \left(\frac{r_m}{r}\right)^8 + C10 \left(\frac{r_m}{r}\right)^{10} \right] \cdot F(r,D) \right\}, \quad (1.2)$$

where $F(r, D)$ is

$$F(r, D) = \begin{cases} e^{-\left(\frac{D\, r_m}{r} - 1\right)^2} & r < D \cdot r_m \\ 1 & r \geq D \cdot r_m \, . \end{cases}$$

The parameters are listed in Table 1.1 and the potential is shown in Fig. 1.3. The minimum is at an internuclear distance of 2.97 Å and has a depth of approximately 10.96 K.

Helium is the element with the smallest electronic polarizability ($\alpha = 0.123\,\text{cm}^3/\text{mol}$ [21]) and therefore the dipole-dipole interaction is very weak, explaining the shallow potential well in the He-He potential. The zero point energy of He atoms localized in the potential wells is much larger than the well depth. A

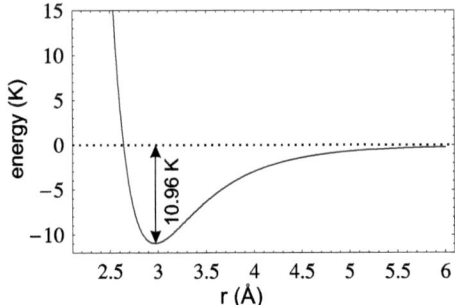

FIGURE 1.3: *He-He interaction potential as a function of the internuclear separation.*

bound structure with a total negative energy can thus only exist if the He-He separation is larger than the distance of the potential wells. One can introduce a parameter $\lambda = E_{kin}/E_{pot}$. Quantum liquids or solids are substances with λ that exceeds unity. In this sense only liquid hydrogen and the two He isotopes are quantum liquids. For He the ratio of zero point energy to the potential energy is about 2.6. It is the reason why He stays liquid down to T=0 K. One can estimate the energy of the zero point oscillation by using the Heisenberg uncertainty relation $\Delta x \cdot \Delta p \sim \hbar$ leading to the expression for the energy

$$\Delta E = \frac{\hbar^2}{2\, M_{He}\, (\Delta x)^2}. \tag{1.3}$$

The solid phase of He can only be reached by applying pressure which forces the atoms to be localized. Solid He is the only quantum crystal in the sense that λ is larger than one. This quantum nature of solid He makes it the softest crystal known. Implanted one electron atoms form small cavities, so called atomic bubbles and they impose their symmetry to the crystal because the Pauli repulsion (host atom-He) exceeds the He-He interaction. In other rare gas matrices guest atoms sit on lattice sites or lattice defects and are strongly perturbed by crystal fields.

1.2.3 Molar volume of solid ^{4}He

Because of its quantum nature condensed helium is very compressible with a compressibility that depends on pressure and temperature. For bubble calculation it is important to know the density ρ_{He} of He or the molar volume $\mathcal{V}_{mol} \propto 1/\rho_{He}$. These two quantities can not be measured directly during the experiment but are inferred from the temperature and the pressure of the crystal.

In this section, analytical forms to calculate the molar volume as a function of temperature and pressure in the bcc and hcp phase of solid He will be given. For liquid He only the relation of pressure to the molar volume at a fixed temperature of 1.6 K is presented. The molar volume depends in general in a much stronger way on the pressure than on the temperature. A detailed discussion of how this analytic forms where obtained can be found in [22]. The basic principle is to measure pressure changes as a function of the temperature at constant volume [23]. In the following all molar volumes are given in cm^3/mol, pressures in bar and the temperature in Kelvin.

In the bcc phase one can assume that the compressibility κ_{bcc} is constant and the molar volume can be expressed as

$$\mathcal{V}_{bcc}(p,T) = \mathcal{V}(T)\Big\{1 - \kappa_{bcc}(p - p_s(T))\Big\}, \tag{1.4}$$

where $p_s(T)$ and $\mathcal{V}_s(T)$ are the pressure and the molar volume at the bcc-liquid phase transition. These two quantities are of the form

$$p_s(T) = p_s^{(1)}(1 + T^{\alpha_s}) + p_s^{(2)} \tag{1.5}$$

$$\mathcal{V}_s(T) = \mathcal{V}_s^{(1)}(1 + T^{\beta_s}) + \mathcal{V}_s^{(2)}.$$

1.2 Properties of liquid and solid ^{4}He

$p_s^{(1)}$ (bar/K)	$p_s^{(2)}$ (bar)	α_s	$\mathcal{V}_s^{(1)}$ ($\frac{mol}{cm^3 K}$)	$\mathcal{V}_s^{(2)}$ ($\frac{mol}{cm^3}$)	β_s	κ_{bcc} (bar^{-1})
0.0506284	25.36	7.98657	-0.00571443	21.196	6.86053	$3.8 \cdot 10^{-3}$

TABLE 1.2: *Parameters used in Eq. 1.4 and Eq. 1.5 for the calculation of the molar volume in the bcc phase of solid He.*

$x_0^{(0)}$ ($\frac{bar}{K^4}$)	$x_1^{(0)}$ ($\frac{bar}{K^5}$)	$x_3^{(0)}$ ($\frac{bar}{K^7}$)	$p^{(0)}$ (bar)
2.44826	-0.998102	1.11305	2341.69
$x_0^{(1)}$ ($\frac{bar\,mol}{K^4\,cm^3}$)	$x_1^{(1)}$ ($\frac{bar\,mol}{K^5\,cm^3}$)	$x_3^{(1)}$ ($\frac{bar\,mol}{K^7\,cm^3}$)	$p^{(1)}$ ($\frac{bar\,mol}{cm^3}$)
-0.277139	0.106835	-0.121036	-211.409
$x_0^{(2)}$ ($\frac{bar\,mol^2}{K^4\,cm^6}$)	$x_1^{(2)}$ ($\frac{bar\,mol^2}{K^5\,cm^6}$)	$x_3^{(2)}$ ($\frac{bar\,mol^2}{K^7\,cm^6}$)	$p^{(2)}$ ($\frac{bar\,mol^2}{cm^6}$)
0.00798785	-0.002849	0.00329108	4.18786

TABLE 1.3: *Parameters used in Eq. 1.7 for the calculation of the molar volume in the hcp phase of solid He.*

The parameters are listed in Table 1.2.

In the hcp phase the corresponding expressions are more complex. A polynomial was fitted to measurements of pressure variation as a function of temperature. This was done for different molar volumes leading to the function

$$p(\mathcal{V},T) = p^{(0)} + \frac{T^4 x_0^{(0)}}{4} + \frac{T^5 x_1^{(0)}}{5} + \frac{T^7 x_3^{(0)}}{7}$$
$$+ (p^{(1)} + \frac{T^4 x_0^{(1)}}{4} + \frac{T^5 x_1^{(1)}}{5} + \frac{T^7 x_3^{(1)}}{7})\mathcal{V}$$
$$+ (p^{(2)} + + \frac{T^4 x_0^{(2)}}{4} + \frac{T^5 x_1^{(2)}}{5} + \frac{T^7 x_3^{(2)}}{7})\mathcal{V}^2. \quad (1.6)$$

Solving Eq. 1.6 for $\mathcal{V}$ gives the molar volume in the hcp phase of solid He as a function of pressure and temperature

$$\mathcal{V}_{hcp}(p,T) = -\frac{140p^{(1)} + 35T^4 x_0^{(1)} + 28T^5 x_1^{(1)} + 20T^7 x_3^{(1)}}{2(140p^{(2)} + 35T^4 x_0^{(2)} + 28T^5 x_1^{(2)} + 20T^7 x_3^{(2)})}$$
$$- \left\{ \frac{(140p^{(1)} + 35T^4 x_0^{(1)} + 28T^5 x_1^{(1)} + 20T^7 x_3^{(1)})^2}{4(140p^{(2)} + 35T^4 x_0^{(2)} + 28T^5 x_1^{(2)} + 20T^7 x_3^{(2)})^2} \right.$$
$$\left. + \frac{140p - 140p^{(0)} - 35T^4 x_0^{(1)} - 28T^5 x_1^{(1)} - 20T^7 x_3^{(1)}}{(140p^{(2)} + 35T^4 x_0^{(2)} + 28T^5 x_1^{(2)} + 20T^7 x_3^{(2)})} \right\}^{1/2}. \quad (1.7)$$

All the parameters are listed in Table 1.3.

In the liquid phase at 1.6 K one can fit the experimental points by the function

$$\mathcal{V}_{liquid}(p) = 27.4701 - 0.27527p + 0.003972p^2. \quad (1.8)$$

As mentioned before, the molar volume does not strongly depend on temperature. Therefore Eq. 1.8 is a good approximation for the molar volume of liquid He at 1.5 K, the normal temperature in our experiments.

1.2.4 Optical properties

Our optical spectroscopy experiments were performed in the visible and near IR range of the spectrum where helium is transparent, since the electronic excitation lies in the VUV region. The excitation lines are far away from the visible and IR range, so one can expect an index of refraction close to one. The index of refraction can be calculated using the Clausius-Mosotti equation

$$n_{He} = \sqrt{\epsilon_{He}} = \sqrt{\frac{3 + 8\pi\alpha/\mathcal{V}_{mol}}{3 - 4\pi\alpha/\mathcal{V}_{mol}}}, \tag{1.9}$$

where $\alpha = 0.123$ cm^3/mol is the polarizability of He [21] already seen in Sect. 1.2.2. With a molar volume of $\mathcal{V}_{mol} = 21.1$ cm^3/mol (T=1.5 K, p=26.8 bar), one gets an index of refraction of 1.0369. At the liquid to solid phase transition the index of refraction changes by 3% due to a 10% change of the He density. At the bcc-hcp phase boundary this change is much smaller ($\approx 0.03\%$). These two phase transitions can therefore be seen by eye. The hcp phase has an axis and is birefringent. However, in our experiment that phase is not mono-crystalline but is rather composed of many small randomly oriented crystals, so that no net birefringent effects can be observed.

1.3 Experimental setup

Matrix isolation spectroscopy of alkali atoms and molecules in solid He, requires low temperatures and high pressures. This demands special experimental efforts and may be one of the reasons why we are the only group doing this kind of research, that has attracted a lot of interest in past years. The research is closely related to spectroscopic studies of atoms on He nanodroplets [24, 25] and in liquid He [26]. The experimental study of the quadratic Stark effect called for highly uniform electric fields in the doped region of the crystal another technical difficulty. An extensive description of technical details of the experimental apparatus that has grown for more than 10 years can be found in earlier works [22, 27]. Here we will present an overview of the cryostat, the pressure cell and the different setups used for the various experimental studies.

1.3.1 The bath cryostat

Figure 1.4 shows a section through the bath cryostat needed to reach the temperature of 1.5 K. The He bath can be filled with 40 l of liquid He. The pressure cell, mounted on an aluminium plate, is immersed in the liquid He. Three pairs of superconducting Helmholtz coils are mounted outside the cell to produce a static magnetic field. Another pair of Helmoltz coils inside the pressure cell is used to apply an oscillating rf-magnetic field. The bath is thermally shielded from ambient temperature by two layers of isolation vacuum pumped by a rotary vane pump (SD-451, Varian) and a turbomolecular pump (TMU 261, Pfeiffer), and pressures of 10^{-7} mbar after filling liquid He in the He bath (cryopump) can be reached. Between the two vacuum layers a chamber filled with liquid nitrogen is used as an additional shield from thermal radiation (shown in Fig. 1.5). Thermal radiation from the top flange is shielded by four gold coated copper plates (baffles) spaced by 10 cm. The cryostat has five windows in three orthogonal directions providing optical access. Figure 1.5 is a schematic cross section through the cryostat showing the pressure cell immersed in the He bath. The capillary mounted on top of the cell is used to admit very pure He gas from an external bottle (of 200 bar) to the cell in order to grow the He crystal. The pressure is measured at the room temperature end of the capillary outside the cryostat. The height adjustable lens above the cell is used to focus radiation of a frequency doubled pulsed Nd:YAG laser onto the alkali metal target mounted at the bottom of the cell, for the implantation of alkali atoms into the He matrix. This process is explained in more detail in Sect. 1.3.3. The cryostat is shielded from laboratory stray magnetic fields by a three layer μ metal shield, not visible in Fig. 1.4. Special care was taken in the selection of the materials used inside the cryostat to reduce stray magnetic fields. Only non-magnetic material such as oxygen free copper, brass or aluminium were used. The materials should not be superconducting at 1.5 K, in order not to trap magnetic flux lines.

A rotary vane pump (Trivac B 65, Leybold) and a roots pump (EHC250, Boc Edwards) are used to cool the liquid He down to 1.5 Kelvin by pumping on the He bath (vapor pressure of 5 mbar). The

1.3 Experimental setup

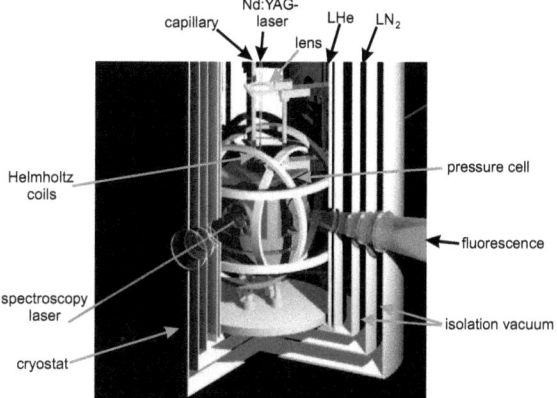

FIGURE 1.4: *Section through the bath cryostat.*

cooling process uses approximately 30% (10-15 liters) of the initial volume of liquid He. The duration of the experiment varies between 48-70 hours depending on the type of experiment. Experiments with electric fields need large HV cables with a relatively large heat conduction to traverse the He bath. This reduces the time of the experiment by a factor 1.5. The temperature is measured inside and outside of the pressure cell by calibrated germanium resistors. The pumping power and thus the temperature are controlled by a electronically controllable butterfly valve on the pumping line.

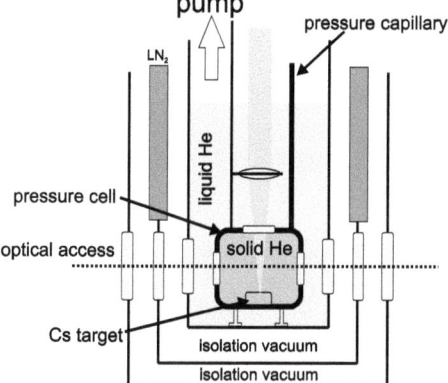

FIGURE 1.5: *Cross section through the cryostat. The height adjustable lens mounted above the cell is used for the implantation process.*

1.3.2 The pressure cell

Helium solidifies only under pressures in excess of 26.5 bars at 1.5 Kelvin. The pressure cell is made of (non-magnetic) copper with an inner volume of 170 cm^3 and has five quartz windows for optical access. It is designed to sustain pressures of 50 bar. The windows mounted on the outside of the cell are sealed with very pure aluminum rings. Electric feedthroughs for the temperature sensor and the rf-coil are mounted on the bottom plate of the cell. A modified version of the pressure cell has two high voltage feedthroughs on the bottom plate and one electric feedthrough for the temperature sensor on the top of the cell. A capillary for applying pressure and the height adjustable lens are also mounted on the top part.

Inside the cell a glass ampoule containing the solid Cs or Rb bulk metal is mounted in a tube on the bottom plate. The ampoule containing the very reactive alkali metals is broken and transferred into the cell under argon atmosphere before the experiment.

For magnetic resonance experiments with static electric fields, glass electrodes connected to the high voltage feedthroughs and rf-Helmholtz coils are mounted on a polycarbonat body inside the cell. Details are explained in [27] and in chapter 8.

1.3.3 Implantation process

Once the He bath is at 1.5 K we grow a He crystal inside the pressure cell by condensing He gas in the cell, and then pressurizing the superfluid He. By controlling the temperature and the pressure in the cell it is possible to produce either the bcc phase or the hcp phase of the He crystal. By solidifying the He at a sufficiently low speed one may produce solid He which consists of large volume single crystals. However, the crystal is locally molten during the implantation process and the resolidification results in a poly crystalline structure. In this sense the hcp phase is not a uniaxial mono-crystal, but consists of small crystals with random orientations. Nevertheless, each atom experiences a local uniaxial perturbation. The implantation is usually done in the harder hcp phase at pressure around 30 bar. For the implantation we use the technique of laser ablation. Three different stages of the implantation process are shown in Fig. 1.6 and are explained below.

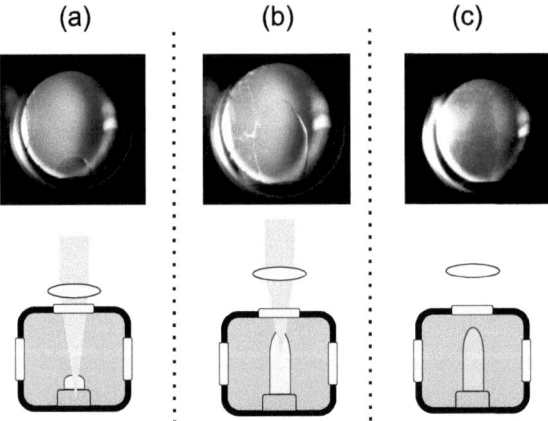

FIGURE 1.6: *Three different stages of the implantation process. The top row shows photographs taken through one of the side windows. (a) strong Nd:YAG pulses are focused on the metallic target. (b) The lens is slowly lifted and the impurities diffuse towards the center of the pressure cell. (c) The laser is switched off, the He resolidifies and the atoms become trapped. The bottom row shows schematic pictures of the process.*

(a) Pulses from a frequency doubled Nd:YAG laser ($\lambda = 532$ nm) with repetition rate of 2-4 Hz and energies of 20-30 mJ/pulse are focused onto the solid metal target (solid Rb, solid Cs or a solid Rb/Cs mixture in a glass tube) mounted on the bottom of the pressure cell. The absorbed heat melts the crystal above the target and at the same time material (atoms, dimers, clusters, ions and electrons) is ablated from the target surface. Locally the ablation produces a plasma creating ions

1.3 Experimental setup

and electrons. The wavelength of the ablation laser is not critical, it is the intensity at the metal surface which must be larger than a certain threshold. We also succeeded to implant atoms by using the signal beam of an OPO (optical parametric oscillator) at around 670 nm (15 mJ). Pure Cs, pure Rb and also 50/50 mixtures of Rb and Cs where successfully implanted by this technique.

(b) After a few hundred pulses the focus of the laser beam is lifted by changing the height of the lens above the cell. The molten part of the crystal moves upwards and the implanted impurities diffuse into the molten part. The crystal melts preferably in the region which contains impurities since the latter absorb the light while the pure He crystal is transparent. In this way the atoms and molecules can be brought into the center of the pressure cell.

(c) When the Nd:YAG laser is switched off, the helium resolidifies and traps the implanted impurities. Atomic densities of $10^8 - 10^9$ cm^{-3} can be achieved in this way. Atoms are mainly lost due to cluster formation [28]. Low energy pulses of the same Nd:YAG laser used for the implantation, are used for the dissociation of those clusters in order to keep the atomic density constant. For this the height of the lens above the cell is adjusted so that the laser beam is focussed in the doped region.

Depending on the implanted alkali metal the doped part of the crystal looks greyish or bluish under white light illumination. This is due to strong absorption bands of metallic clusters peaked at around 760 nm (in case of a Rb doped crystal). One of our fascinating observations is, that this blue column stays solid below the melting point of pure He. Details of this feature are presented in chapter 3.

References

[1] M. Arndt, S. I. Kanorsky, A. Weis, and T. W. Hänsch, Phys. Lett. A **174**, 298 (1993).
[2] A. Weis, S. Kanorsky, S. Lang, and T. W. Hänsch, in *Lecture Notes in Physics* (Springer, 1997).
[3] S. Kanorsky, A. Weis, M. Arndt, R. Dziewior, and T. Hänsch, Z. Phys. B **98**, 371 (1995).
[4] S. I. Kanorsky, M. Arndt, R. Dziewior, A. Weis, and T. W. Hänsch, Phys. Rev. B **50**, 6296 (1994).
[5] T. Kinoshita, Y. Takahashi, and T. Yabuzaki, Phys. Rev. B **49**, 3648 (1994).
[6] S. I. Kanorsky, M. Arndt, R. Dziewior, A. Weis, and T. W. Hänsch, Phys. Rev. B **49**, 3645 (1994).
[7] M. Arndt, S. I. Kanorsky, A. Weis, and T. W. Hänsch, Phys. Rev. Lett **74**, 1359 (1995).
[8] S. I. Kanorsky, S. Lang, S. Lücke, S. B. Ross, T. W. Hänsch, and A. Weis, Phy. Rev. A **54**, R1010 (1996).
[9] P. Moroshkin, A. Hofer, S. Ulzega, and A. Weis, Fiz. Nizk. Temp. **32**, 1297 (2006), (Low Temp. Phys. 32(11), 981-998 (2006)).
[10] H. K. Onnes, Nature **77**, 559 (1908).
[11] W. H. Keesom, Nature **118**, 81 (1926).
[12] P. Kapitza, Nature **141**, 74 (1938).
[13] J. F. Allen and A. D. Misener, Nature **141**, 75 (1938).
[14] W. H. Keesom and K. W. Taconis, Physica **5**, 161 (1938).
[15] J. S. Dugdale and F. E. Simon, Proc. Royal Soc. London. Series A **218**, 291 (1953).
[16] J. H. Vignos and H. A. Fairbank, Phys. Rev. Lett. **6**, 265 (1961).
[17] A. F. Schuch and R. L. Mills, Phys. Rev. Lett. **8**, 469 (1962).
[18] E. B. Osgood, V. J. Minkiewicz, T. A. Kitchens, and G. Shirane, Phys. Rev. A **5**, 1537 (1972).
[19] V. J. Minkiewicz, T. A. Kitchens, F. P. Lipschultz, R. Nathans, and G. Shirane, Phys. Rev. **174**, 267 (1968).
[20] R. A. Aziz and A. R. Janzen, Phys. Rev. Lett. **74**, 1586 (1995).
[21] A. K. Bhatia and R. J. Drachman, J. Phys. B **27**, 1299 (1994).
[22] R. Müller-Siebert, Ph.D. thesis, University of Fribourg, Switzerland (2002).
[23] J. F. Jarvis, D. Ramm, and H. Meyer, Phys. Rev. **170**, 320 (1968).
[24] F. R. Brühl, R. A. Trasca, and W. E. Ernst, J. Chem. Phys. **115**, 10220 (2001).
[25] O. Bünermann, M. Mudrich, M. Weidemüller, and F. Stienkemeier, J. Chem. Phys. **121**, 8880 (2004).
[26] T. Kinoshita, K. Fukuda, Y. Takahashi, and T. Yabuzaki, Phys. Rev. A **52**, 2707 (1995).
[27] S. Ulzega, Ph.D. thesis, University of Fribourg, Switzerland (2006).
[28] S. Lang, Ph.D. thesis, Ludwig Maximilians Universität München (1997).

Part II

Optical spectroscopy of atoms and exciplexes in solid ^{4}He

Chapter 2

Introduction to spectroscopic and time-resolved measurements in solid ^{4}He

2.1 Optical spectroscopy

The matrix isolation technique is used to investigate atomic and molecular impurities trapped in a chemically inert solid matrix by different means. This technique has been known since the 1950s. Rare gas matrices are ideal hosts because no chemical reaction takes place between the impurities and the rare gas atoms. But in heavy rare gas matrices (like Ne, Ar, Kr, Xe) the impurities are strongly perturbed, making the observations difficult to interpret. The impurities reside on regular or substitutional lattice sites. Their properties are therefore strongly affected by the symmetries of the local trapping sites.

Solid He is special due to the weak He-He interaction which can be overcome easily by the interaction between the impurity atom and the He atoms. The size and shape of the trapping site is mainly determined by the impurity-He interaction. The impurity repels the He atoms leading to the formation of a cavity around the guest atom shown in Fig. 2.1. The existence of such cavities was first suggested for electrons in liquid He [1] and the structures were called electron bubbles. The size of the bubble is determined by the balance of the repulsive electron-He interaction and the energy needed to form the bubble. The bubble model developed for electrons in liquid He [2] and later extended for atomic impurities in liquid He [3, 4], describes liquid helium as an incompressible continuous medium. This model can be extended to solid He [5], in which the strong delocalization of the constituent atoms and the large overlap of their wavefunctions justifies the description of this quantum solid as a continuous medium. The question of compressibility is addressed in chapter 4.

Alkali atoms embedded in liquid He and the bcc phase of solid He impose their symmetry on the bubble to a large extend. The $n\,S_{1/2}$ and $n\,P_{1/2}$ ground and first excited states are spherically symmetric and thus the corresponding bubble is spherical. The bubble model explains the shift and broadening of optical lines of alkali atoms embedded in liquid and solid He [6, 5]. An extended bubble model (presented in chapter 4) allowed us to calculate the wavefunction of Cs atoms in liquid and solid He and explaining the jumps of the excitation and emission lines at the liquid-solid phase boundary.

The setup for optical spectroscopic measurements is shown in Fig. 2.2. The defects (atoms or molecules) trapped in the He matrix are excited by different lasers covering the broad spectral range from 430 nm up to 950 nm. The fluorescence light is collimated by a lens and then focused into a grating spectrograph (Oriel MS257) for spectrally resolved detection. For the detection we use either a CCD camera, different types of photodiodes or two different photomulitipliers mounted on the output port of the spectrograph, depending on the spectral range and the response time needed for the specific experiment. Fast detectors (photomultiplier and small area Si-photodiodes) are used for time-resolved measurements.

Lifetime measurements of the Cs $6P_{1/2}$ state were performed using time-resolved photon counting. The setup for this experiments including results is presented in chapter 5.

A pulsed detection system was developed for the recording of spectra using a pulsed excitation laser in combination with a photodiode or a photomultiplier. For those measurements the grating of the spectrograph is rotated step by step and at each position an average over several fluorescence pulse shapes is recorded with an oscilloscope and the calculated area of the recorded peak is read out by a PC.

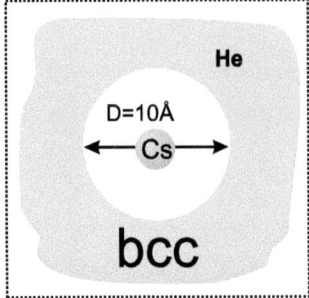 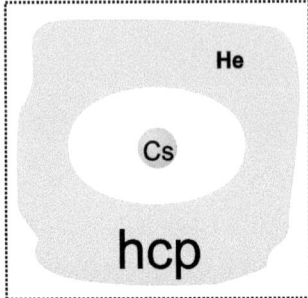

FIGURE 2.1: *Sketch of a Cs ground state atom in cubic (bcc) and hexagonal (hcp) solid He. The Pauli repulsion leads to the formation of an atomic bubble. The symmetry of the bubble (spherical in bcc, deformed sphere in hcp) reflects the symmetry of the atom and the (an-) isotropy of the crystal. The He bulk can be regarded as a continuous medium due to its quantum nature.*

With this system very low signal levels can be detected and time-resolved measurements can be done by measuring the evolution of the fluorescence pulse.

2.1.1 Atomic bubble and optical absorption-emission cycle

The interpretation of optical lines using the bubble model uses the fact that the optical excitation and emission processes can be treated independently (Fig. 2.3) [3]. The absorption-emission cycle of the D_1 line ($6S_{1/2} \to 6P_{1/2}$) of Cs in solid He consists of the following discrete steps:

(**1**) First the Cs atom is in the $6S_{1/2}$ ground state and resides in an equilibrium bubble with a diameter of around 12 Å. According to the Frank-Condon principle, the shape and size of the bubble does not change during the fast laser excitation (10^{-13} s) to the $6P_{1/2}$ excited state. The wavefunction of the electron in the excited state extends to larger distances from the atomic core than the ground state wavefunction. It is more perturbed by the interaction with the bubble than the ground state wavefunction and therefore the excitation line shows a relatively large blue shift with respect to the free atomic excitation line.

(**2**) After the excitation process the bubble relaxes, thereby minimizing the total energy of the bubble. This relaxation takes place on a timescale of picoseconds, much faster than the radiative lifetime of the $6P_{1/2}$ excited state, which is on the order of 30 ns (presented in chapter 5).

(**3**) The spontaneous emission to the ground state takes place in the relaxed bubble. The equilibrium bubble of the $6P_{1/2}$ state has a diameter of 15 Å. It is larger than the ground state bubble, explaining the smaller blue shift of the emission line compared to the excitation line.

(**4**) After the emission the Cs atom is again in the $6S_{1/2}$ ground state and the bubble shrinks to its initial equilibrium size of the ground state.

2.2 Alkali dimers in solid He

Optical spectroscopic studies of Cs and Rb atoms in solid He were done by our group for many years. It is only very recently that we investigated alkali dimer molecules in a solid He matrix. This section gives a short overview of what was done in this field during the time of my thesis.

We have observed several absorption bands of the homo-nuclear Cs_2, Rb_2 and the hetero-nuclear CsRb dimers, in which only one molecular emission band for each dimer could be observed. The emission

2.2 Alkali dimers in solid He

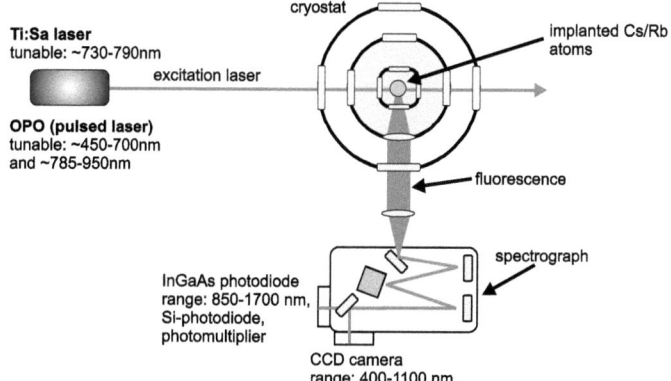

FIGURE 2.2: *Setup for the spectroscopic measurements.*

occurs always from the lowest excited triplet state to the singlet ground state (forbidden transition). In the following we will focus on the Rb_2 dimer [7]. The other two dimers show similar features.

The experimental setup is the same as for the spectroscopy of alkali atoms shown in Fig. 2.2. The He crystal is doped by laser ablation explained in Sect. 1.3.3. As excitation laser we use the pulsed signal or idler beams of the OPO and the continuous beam from a tunable Ti:Sa laser. In this way we can cover a very broad spectral range, indicated in Fig. 2.2. Some experiments were done by sending the signal or idler beam of the OPO from top of the cryostat, along the path of the ablation laser. In this way the excitation laser passes the implanted region along its more extended vertical direction and is focused somewhere in the center of it, resulting in general in larger signals.

Figure 2.4 shows a typical fluorescence spectrum recorded with a CCD camera after excitation on a Rb_2 absorption band (from a Rb doped He crystal). Three different features can be observed in the spectrum: i) emission on the D_1 and D_2 line of atomic Rb at 780 nm, produced via photodissociation of the Rb_2 dimer ii) emission from Rb^*He_2 exciplexes at 850 nm and iii) the emission from the Rb_2 dimer at 1042 nm. The exciplex emission will be presented in detail in chapter 6. The emission at 1042 nm is the only molecular emission found in the spectral range from 500-1600 nm, when the excitation wavelength is tuned from 450-900 nm.

We have identified 10 different excitation bands, leading either to molecular emission at 1042 nm or photodissociation followed by atomic and exciplex emission or both together. We have done model calculation using molecular potentials and were able to identify all observed absorption bands. Details can be found in [7]. Figure 2.5 shows the theoretical *ab initio* potential curves from Ref. [8]. We calculated the vibrational states and corresponding wavefunctions of the molecule by solving the one dimensional Schrödinger equation. The electronic transitions were evaluated by calculating the overlap integrals between the excited and ground state wavefunctions.

In addition we did time-resolved measurements of the molecular fluorescence at 1042 nm. A characteristic fluorescence pulse shape is shown in Fig. 2.6 together with a pulse of the scattered excitation laser. The scattered laser pulse (FWHM of 3 μs) reflects the time resolution of the detection system. The decay time of the fluorescence at 1042 nm is well resolved and has an exponential decay with a time constant of 50 μs. Excitation at different wavelength does not change the decay time. In addition one can also see a finite rise time of the fluorescence.

Our interpretation of the slow rise time is as follows. After laser excitation to some excited state of the molecule, no molecular emission except for the emission from the lowest excited state can be observed. The quenching of all upper excited states down to the lowest excited state, via radiationless decay or far infrared emission which we can not detect, seems to be faster than the radiative lifetime of these states. The slow onset of the molecular emission however shows that the quenching involves a metastable

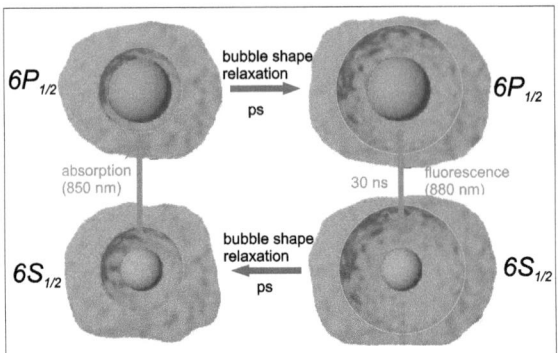

FIGURE 2.3: *The four steps of the optical absorption-emission cycle. Details are given in the text.*

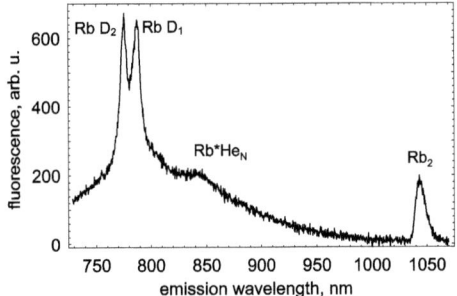

FIGURE 2.4: *Emission spectrum from a Rb doped He crystal (T=1.5 K, P=31 bar). Excitation by the signal beam of the OPO at 570 nm.*

state which must be the lowest excited state $^3\Pi_u$. The population of this state is followed by the slow formation of a molecular exciplex $Rb_2^*He_n$ with subsequent fluorescence. The close He atoms forming the exciplex are arranged on a ring around the waist of the dumbell-like electronic orbit of the excited $^3\Pi_u$ state. These He atoms perturb the molecule and lift the selection rule that forbids its radiative transition to the singlet ground state.

In the following we summarize the main characteristics for the three observed dimers.

Rb$_2$: emission at 1042 nm, lifetime of 50 µs.

Cs$_2$: emission at 1160 nm, lifetime of 40 µs.

CsRb: emission at 1240 nm, lifetime of 45 µs.

2.2 Alkali dimers in solid He

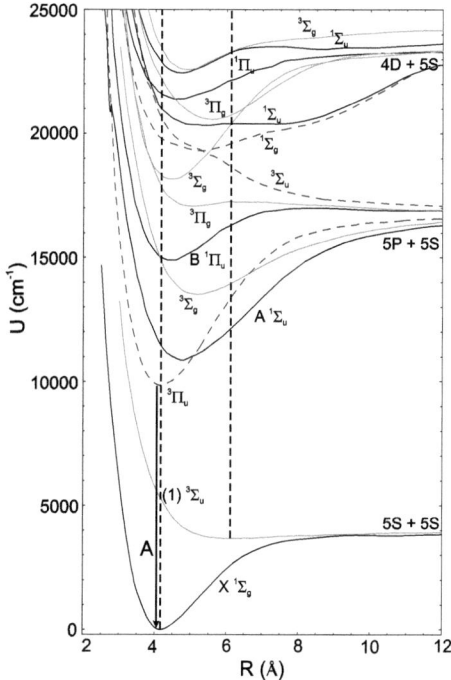

FIGURE 2.5: *Potential diagram of the Rb_2 system. Dashed potential lines have no allowed transition to the ground state. The vertical dashed lines indicate the excitations form the singlet and the triplet ground state. The observed Rb_2 emission at 1042 nm is indicated by an arrow with label A.*

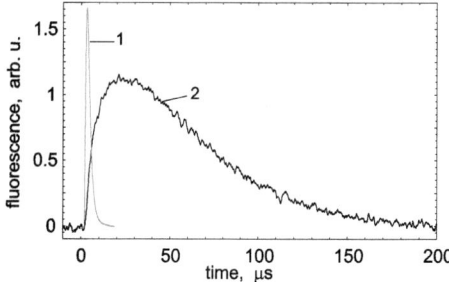

FIGURE 2.6: *Pulse shape of the Rb_2 fluorescence at 1042 nm (black line (2)). The red line (1) shows the recorded pulse of scattered laser light which determines the time resolution of the detection system (InGaAs photodiode plus current amplifier).*

References

[1] R. A. Ferrell, Phys. Rev. **108**, 167 (1957).
[2] J. Jortner, W. R. Kestner, S. A. Rice, and M. H. Cohen, J. Chem. Phys. **43**, 2614 (1965).
[3] S. I. Kanorsky, M. Arndt, R. Dziewior, A. Weis, and T. W. Hänsch, Phys. Rev. B **50**, 6296 (1994).
[4] T. Kinoshita, K. Fukuda, Y. Takahashi, and T. Yabuzaki, Phys. Rev. A **52**, 2707 (1995).
[5] S. Kanorsky, A. Weis, M. Arndt, R. Dziewior, and T. Hänsch, Z. Phys. B **98**, 371 (1995).
[6] A. P. Hickman, W. Steets, and N. F. Lane, Phys. Rev. B **12**, 3705 (1975).
[7] P. Moroshkin, A. Hofer, S. Ulzega, and A. Weis, J. Chem. Phys. **74**, 032504 (2006).
[8] S. J. Park, S. W. Suh, Y. S. Lee, and G. H. Jeung, J. Mol. Spect. **207**, 129 (2001).

Chapter 3

Paper I:
Impurity-stabilized solid ^{4}He below the solidification pressure of pure helium.

This paper reports on our studies of the solid structure (iceberg) doped with alkali atoms, clusters and charged particles that appears after the melting of the He crystal.

My main contributions to the work were:

- Setting up the interferometer and the extinction spectrometer and performing the extinction and interferometric measurements of the iceberg together with P. Moroshkin.
- Interpretation of the observations in discussions with P. Moroshkin and A. Weis.
- Producing figures and text for the paper.

Impurity-stabilized solid ^{4}He below the solidification pressure of pure helium.

P. Moroshkin[1], A. Hofer[1], S. Ulzega[2] and A. Weis[1]

[1]*Département de Physique, Université de Fribourg, Chemin du Musée 3, 1700 Fribourg, Switzerland*
[2]*EPFL, Lausanne, Switzerland.*

Published in Nature Physics **3**, 786 (2007).

Abstract: The modification of melting temperatures and pressures by dissolved impurities is well known in classical fluids. To our knowledge such effects have never been studied in quantum solids because of the difficulties in introducing impurities into such crystals which exist only at cryogenic temperatures, and, in the case of ^{4}He, at pressures exceeding 25 bar. Here we present a dramatic effect that occurs during the melting of solid ^{4}He doped with nanoscopic impurities — alkali atoms, clusters, ions, and electrons: the doped part of the crystal remains solid under conditions at which pure helium is liquid. Using interferometry we found that the density of the solid structure (iceberg) lies between the densities of pure liquid and pure solid helium. We tentatively interpret the iceberg as being an aggregation of positively charged particles and electron bubbles.

The coexistence of a ^{4}He crystal with superfluid ^{4}He is a model system for investigating fundamental aspects of the growth and melting of quantum crystals. Many subtle effects at the crystal-liquid interface have been studied in the past, such as faceting and propagation of crystallization waves (reviewed in [1]), interface motion under the action of acoustic waves [2] or confined electrons [3], instability induced by a nonhydrostatic stress [4]. The role of ^{3}He impurities was discussed in connection with the formation of supersolid ^{4}He [5, 6].

In the past decade we have developed a laser ablation technique for doping a ^{4}He crystal with atomic impurities. Our (recently reviewed [7]) main research activity is devoted to optical and magnetic resonance spectroscopy of atomic, molecular, and exciplex defects, both in bcc (body-centered cubic) and in hcp (hexagonal close-packed) ^{4}He crystals. The doped part of the crystal has approximately the shape of a vertical cylindrical column of bluish color, which can be observed through a side window of the cell (Figs. 3.1a–d).

When the crystal pressure is slowly decreased, an interesting phenomenon is observed during the solid-liquid phase transition. Figs. 3.1a and 3.1b show the doped crystal in the hcp and bcc phases respectively. While the hcp phase is perfectly homogeneous and transparent, the bcc phase has a polycrystalline structure and light refraction at the grain boundaries gives the sample a turbid, albeit not completely opaque, appearance (Fig. 3.1b). The crystal melts from top to bottom when the pressure is further released. The well visible liquid-solid phase boundary moves downwards when the pressure is lowered. In Fig. 3.1c, the liquid level has dropped below the upper end of the doped column. Most remarkably, the doped column remains a solid and stable structure protruding into the liquid until all surrounding helium is liquefied (Fig. 3.1d), while still under pressure. Under those conditions the structure remains unchanged for at least half an hour. A further relief of pressure makes the structure break up into many smaller parts which then float to the bottom of the pressure cell (Fig. 3.2). We will refer to the solid structure in the liquid matrix as an "iceberg".

We have observed similar structures in crystals doped by ablation from targets of cesium, rubidium, or their 50/50 mixture which all behave similarly. In order to better understand the nature of the icebergs, we have carried out the investigation reported below.

The coloring of the sample (Fig. 3.1) is due to a broad absorption band centered around 750 nm

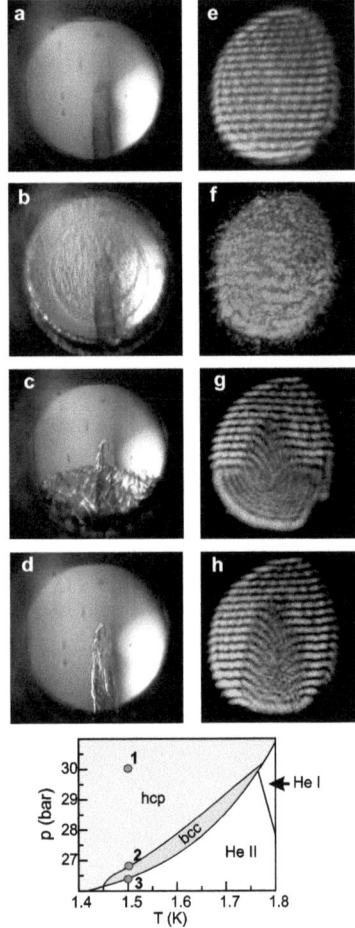

FIGURE 3.1: **Appearance of iceberg structure during melting of a doped ^{4}He crystal.** Photographs of Rb (a–d) and interferograms of Cs (e–h) doped He taken through a 2 cm diameter window during controlled pressure release at T = 1.5 K. Corresponding experimental conditions are shown in the phase diagram as dots. (a,e @ 1): hcp crystalline phase at 30 bar, (b,f @ 2): transition from hcp to bcc at 26.8 bar, (c,g @ 3): bcc–liquid phase transition at 26.4 bar, (d,h @ 3) liquid He at 26.4 bar, just at the end of the phase transition. The column-like structure in the center (iceberg) corresponds to the part doped with Cs/Rb.

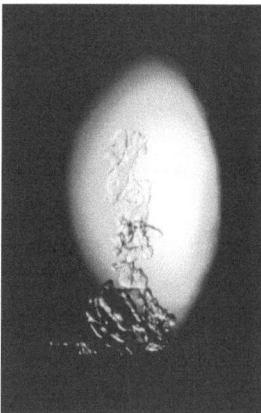

FIGURE 3.2: **Cesium doped iceberg in liquid helium during its disintegration.** The shape of the illuminated region results from the oblique view through two aligned windows. T=1.5 K, p=26.4 bar

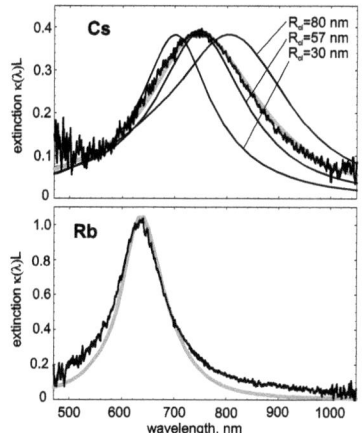

FIGURE 3.3: **Extinction spectra of Cs and Rb doped solid He.** The noisy curves are experimental data and the thick line is the result from Mie theory with a Gaussian distribution of cluster sizes as discussed in the text. For Cs we also show the Mie results for monodisperse clusters (thin lines).

for Cs and 640 nm for Rb (Fig. 3.3) that is typical for plasmon resonances in metallic nano-particles [8]. The spectral dependence of the extinction coefficient depends on the cluster size and on the complex dielectric function $\epsilon(\lambda)$ of the metal. It is described by Mie theory [9] for small particle sizes ($R_{cl} \ll \lambda$). Using dielectric functions from [10] we have modeled the extinction spectra by assuming spherically-shaped clusters and fitted the experimental spectra by adjusting the radii R_{cl} and density distribution $N_{cl}(R_{cl})$ of the Rb_n and Cs_n clusters (Fig. 3.3). For cesium, a best fit with the experimental spectrum is obtained for an average cluster radius of $\langle R_{cl} \rangle = 41$ nm, a width of the distribution (FWHM) of $\Delta R_{cl} = 50$ nm and a total cluster density $N_{cl} = 3.3 \times 10^9$ cm^{-3} (assuming a 3 mm diameter of the doped part of the crystal). The corresponding parameters for rubidium are $\langle R_{cl} \rangle = 10$ nm, $\Delta R_{cl} = 41$ nm and $N_{cl} = 2.5 \times 10^{10}$ cm^{-3}.

We have determined the density of the doped sample from the index of refraction measured with a Mach–Zehnder interferometer. The index of refraction of pure solid helium is $n_{He} = \sqrt{1 + 4\pi\alpha_{He}(\lambda)\rho} = 1.03716$ for a molar polarizability $\alpha_{He}(\lambda) = 0.125$ cm^3/mol [11] (at $\lambda = 546$ nm) and a molar density $\rho = 4.82 \times 10^{-2}$ mol/cm^3 (at $p = 30$ bar and $T = 1.5$ K). The index of refraction n_{cl} of nano-particles suspended in a dielectric matrix with index of refraction n_{He} can be calculated from Mie theory. One can show — using N_{cl} and R_{cl} inferred from the extinction measurement — that the clusters change the index of refraction by $\Delta n_{cl} = 2 \times 10^{-6}$ which induces an additional phase shift of only $0.01 \times 2\pi$ and hence does not affect the fringe pattern compared to the pattern observed with pure helium.

The left and right columns of Fig. 3.1 were obtained with different samples (Rb on the left, Cs on the right), but corresponding pictures in each row were recorded during similar phases of the melting process. Interferograms taken in doped solid He are indistinguishable from those taken in pure liquid or pure solid helium and reveal no structure that could be associated with the dopants (atomic or cluster). A typical fringe pattern obtained with Cs doped (hcp) solid He is shown in Fig. 3.1e.

Lowering the He pressure in the cell results in a uniform upward motion of the fringe pattern. During the phase transition (hcp to bcc, and also bcc to liquid) the sample becomes inhomogeneous, due to the coexistence of the two involved phases (Fig. 3.1b and lower part of Fig. 3.1c) and the fringes disappear (Fig. 3.1f and lower part of Fig. 3.1g).

When the pressure is further released, the crystal starts to melt and the two phases separate: the solid occupies the lower part of the cell, and the liquid the upper part, as shown in Figs. 3.1c, 3.1g. The iceberg structure rises from the bcc phase into the He liquid (which is homogeneously transparent) yielding a horizontal fringe pattern, while the fringes are deformed in the region of the iceberg. Finally, Figs. 3.1d and 3.1h show the iceberg after the surrounding helium is completely liquefied.

The fringes in the region of the iceberg are strongly deformed, indicating that the refractive index of the iceberg differs from that of the surrounding liquid helium. The curvature of the fringes is related to the variation of the thickness of the (approximately cylindrical) iceberg across the picture, and the fringe shift (with respect to its position in the liquid) reaches a maximum in the center of the iceberg, where it is thickest. As discussed above, the dopants (atoms and/or clusters) have a negligible contribution to the refractive index which is thus determined only by the He density. From the downward bending of the fringes and their upward motion during pressure release we conclude that the iceberg density is larger than that of liquid helium.

The fringes produced by a ≈ 3 mm thick slab of helium (the iceberg diameter) move by approximately 16 fringe periods when going from solid to liquid, as is easily estimated based on the refractive indices of helium given above. From Figs. 3.1g and 3.1h one can estimate a maximal deformation of the pattern to be 3 to 4 fringe periods in the center of the iceberg. This leads us to conclude that the density of the iceberg lies somewhere between the densities of liquid and solid helium, although closer to the liquid density.

The solid structure in cesium doped helium reported here has several common features with the macroscopic solid structures, called impurity-helium solids, investigated in [12, 13, 14, 15]. In those experiments, a helium gas jet doped with molecular or atomic impurities was directed into liquid helium. Upon condensation in the helium bath, the gas mixture forms a macroscopic highly-porous structure composed of frozen impurity clusters surrounded by a relatively thin shell of solid helium [14, 15]. According to a theoretical model presented in [12] the solidification of He in those structures, i.e., the attachment of He atoms to the impurity center, is due to the van der Waals attraction between the impurity atoms or molecules (which have *paired* electrons) and the surrounding helium atoms.

The structure observed here has a different underlying binding mechanism. The strong repulsion (due to the Pauli principle) between the *unpaired* Cs/Rb valence electron and the closed S-shell of a He atom dominates over the attractive van der Waals force [16]. Moreover, the estimated [17] number density of Cs atoms (10^8–10^9 cm^{-3}) is many orders of magnitude smaller than the dopant concentrations used in the experiments with impurity-helium solids (about 10^{20} cm^{-3}). Alkali clusters also have a very low number density, and the electron density on their surface is so high that He atoms cannot approach close enough to experience a van der Waals attraction [18], as evidenced by the non-wetting of solid Cs by superfluid ^{4}He [19]. It is therefore very unlikely that *neutral* particles can hold together such a large amount of He atoms and the observed iceberg structure must be due to another type of impurities produced during ablation. These impurities are not detected by our spectroscopic (extinction, laser induced fluorescence [7]) and interferometric measurements.

Based on these facts we assume that charged particles are responsible for the iceberg formation. It is well known [20, 21] that alkali ions in liquid He attract He atoms via electrostriction and form so-called snowballs — complexes consisting of one alkali ion surrounded by a spherical shell of He atoms whose local density is so high that it is solid. According to recent theoretical studies [22, 23] up to 17 He atoms can be bound to a single Cs$^+$ ion in liquid He. It is also well known that the electrostriction produced by an externally applied inhomogeneous electric field in liquid helium at appropriate temperature and pressure can initiate He crystal nucleation [24]. Both atomic and cluster ions can be produced by the laser ablation from a metallic surface and are thus probably present in our samples. The extinction spectra in Fig. 3.3 contain contributions from neutral and positively charged clusters since the plasmon resonance frequencies of large clusters depend on the number of electrons, N_e and are thus rather insensitive to their degree of ionization. Highly charged clusters break up into smaller, weakly charged clusters by Coulomb explosion. The density of charged clusters is thus smaller than N_{cl} determined from Fig. 3.3. It is likely — although, to our knowledge not studied — that charged clusters also form snowballs like atomic ions do. We therefore believe that atomic (and eventually cluster) ions play a dominant role in binding He atoms together. Unfortunately, the absorption lines of Cs$^+$ and Rb$^+$ ions lie in the deep UV part of the spectrum, not accessible to our spectrometer.

Our observation of electric field induced currents in doped crystals [25] points to the presence of charges. Another manifestation of the presence of charged impurities was observed when the sample (iceberg) is destroyed by releasing He pressure in the presence of an electric field. After the melting of surrounding solid He, fragments of different sizes start to split off the iceberg and to move towards one of the high voltage electrodes, where they stick and stay attached until the complete destruction. At some point the whole iceberg splits into two groups of fragments attracted each by one of the electrodes.

Crystalline structures formed by ionic snowballs were recently observed [26] in experiments with He^+ ions in liquid He which form two-dimensional crystals when confined by electric fields. The iceberg structure reported in the present paper is stable without any external field. We suggest that the repulsive Coulomb interaction between the ions is compensated by the attractive forces due to electrons — other negatively charged particles are unlikely to be present — distributed in the same doped volume. Indeed, the laser ablation produces equal amounts of positive and negative charges and the resulting iceberg as a whole is electrically neutral. The recombination of the charged particles, while very efficient in the gas phase, is strongly suppressed in condensed He due to the stabilizing effect of the snowball shell surrounding each positive ion and the bubble structure of the electron [27]. From the Pauli principle, the electron experiences a strong repulsion by the closed S-shells of the He atoms and cannot come close enough to the snowball core thereby suppressing recombination.

In conclusion we believe that the iceberg consists of the aggregation of positively charged particles and electron bubbles probably assisted by surface tension [28]. Whether these particles form a disordered or an ordered (poly-)crystalline structure is not clear. Future studies using IR spectroscopy of electron bubbles or a measurement of plasma oscillations, e.g., may shine more light on the nature of the iceberg.

Methods

Experiments were performed at 1.5 K over a range of pressures covering 26 to 36 bar. The phase diagram of ^{4}He (Fig. 3.1) shows that at 1.5 K and 26.4 bar, superfluid He solidifies in a bcc crystalline structure which then makes a phase transition to a hcp structure at 26.8 bar.

The implantation technique is described in detail in [17, 7]. A helium crystal is grown inside a pressure cell (inner volume $\approx$ 200 ml) by condensing and then solidifying pressurized helium gas from an external reservoir. The cell is immersed in a helium bath kept at 1.5 K by pumping on the bath. Four lateral and a top window provide optical access from three orthogonal directions. The solid host matrix is doped with Cs/Rb atoms by means of laser ablation with the second harmonic of a pulsed frequency-doubled Nd:YAG-laser focused through the top window onto a solid alkali metal target at the bottom of the cell.

For the extinction measurements a collimated beam of white light from a halogen lamp was sent through the sample and the spectrum $I(\lambda)$ of the transmitted light was analyzed by a grating spectrograph equipped with a CCD camera. The extinction coefficient is defined by $\kappa(\lambda)L = -\ln I(\lambda)/I_0(\lambda)$, where L is the sample thickness, and $I_0(\lambda)$ is a reference spectrum recorded after the complete melting of the crystal.

The index of refraction of the sample is measured using a two-beam Mach–Zehnder interferometer illuminated with a beam from a green laser pointer (532 nm) expanded to the size of the cell windows (2 cm diameter). One of the interferometer arms crosses the sample. A small angle is introduced between the two interfering beams in order two obtain a pattern of 10–20 horizontally-oriented interference fringes covering both the doped and the undoped parts of the matrix. The fringe pattern is projected onto a screen and photographed with a digital camera (Figs. 3.1e – 3.1h).

ACKNOWLEDGMENTS This work was supported by grant No. 200020–103864 of the Schweizerischer Nationalfonds. Correspondence and requests for materials should be addressed to A.W.

Competing financial interests

The authors declare that they have no competing financial interests.

References

[1] S. Balibar, H. Alles, and A. Ya. Parshin. The surface of helium crystals. *Rev. Mod. Phys.*, 77:317–370, 2005.

[2] R. Nomura, S. Kimura, F. Ogasawara, H. Abe, and Y. Okuda. Orientation dependence of interface motion in ^{4}He crystals induced by acoustic waves. *Phys. Rev. B*, 70:054516, 2004.

[3] D. Savignac and P. Leiderer. Charge-induced instability of the ^{4}He solid-superfluid interface. *Phys. Rev. Lett.*, 49:1869–1872, 1982.

[4] R. H. Torii and S. Balibar. Helium crystals

under stress: the Grinfeld instability. *J. Low Temp. Phys.*, 89:391–400, 1992.

[5] E. Kim and M. H. W. Chan. Probable observation of a supersolid helium phase. *Nature*, 427:225–227, 2004.

[6] N. Prokof'ev. What makes a crystal supersolid? *Advances in Physics*, 56(1–2):381–402, 2007.

[7] P. Moroshkin, A. Hofer, S. Ulzega, and A. Weis. Spectroscopy of atomic nd molecular defect in solid ^{4}He using optical, microwave, radio frequency, magnetic and electric fields. *Fiz. Nizk. Temp.*, 32:1297–1319, 2006. (Low Temp. Phys., 32:981-998, 2006).

[8] S. Mochizuki, K. Inozume, and R. Ruppin. Spectroscopic studies of rubidium vapour zone produced by thermal evaporation in noble gas. *J. Phys.: Condens. Matter*, 11:6605–6619, 1999.

[9] H. C. van de Hulst. *Light scattering by small particles*. Dover Publications, New York, 1981.

[10] N. V. Smith. Optical constants of rubidium and cesium from 0.5 to 4.0 eV. *Phys. Rev. B*, 2:2840–2848, 1970.

[11] M. H. Edwards. Refractive index of ^{4}He: Saturated vapor. *Phys. Rev.*, 108:1243–1245, 1957.

[12] E. B. Gordon, V. V. Khmelenko, A. A. Pelmenev, E. A. Popov, O. F. Pugachev, and A. F. Shestakov. Metastable impurity-helium solid phase. experimental and theoretical evidence. *Chem. Phys.*, 170:411–426, 1993.

[13] R. E. Boltnev, E. B. Gordon, V. V. Khmelenko, I. N. Krushinskaya, M. V. Martynenko, A. A. Pelmenev, E. A. Popov, and A. F. Shestakov. Luminescence of nitrogen and neon atoms isolated in solid helium. *Chem. Phys.*, 189:367–382, 1994.

[14] S. I. Kiselev, V. V. Khmelenko, D. M. Lee, V. Kiryukhin, R. E. Boltnev, E. B. Gordon, and B. Keimer. Structural studies of impurity-helium solids. *Phys. Rev. B*, 65:024517, 2001.

[15] E. B. Gordon. Impurity condensation in liquid and solid helium. *Low Temp. Phys.*, 30:756–762, 2004.

[16] J. Pascale. Use of l-dependent pseudopotential in the study of alkali-metal-atom-He systems. the adiabatic molecular potential. *Phys. Rev. A*, 28(2):632, 1983.

[17] M. Arndt, R. Dziewior, S. Kanorsky, A. Weis, and T. W. Hänsch. Implantation and spectroscopy of metal atoms in solid helium. *Z. Phys. B*, 98(3):377–381, 1995.

[18] E. Cheng, M. W. Cole, J. Dupont-Roc, W. F. Saam, and J. Treiner. Novel wetting behavior in quantum films. *Rev. Mod. Phys.*, 65:557–567, 1993.

[19] P. J. Nacher and J. Dupont-Roc. Experimental evidence for nonwetting with superfluid helium. *Phys. Rev. Lett.*, 67:2966–2969, 1991.

[20] M. W. Cole and R. A. Bachman. Structure of positive impurity ions in liquid helium. *Phys. Rev. B*, 15:1388–1394, 1977.

[21] W. I. Glaberson and W. W. Johnson. Impurity ions in liquid helium. *J. Low Temp. Phys.*, 20:313–338, 1975.

[22] M. Buzzacchi, D. E. Galli, and L. Reatto. Alkali ions in superfluid ^{4}He and structure of the snowball. *Phys. Rev. B*, 64:094512, 2001.

[23] M. Rossi, M. Verona, D. E. Galli, and L. Reatto. Alkali and alkali-earth ions in ^{4}He systems. *Phys. Rev. B*, 69:212510, 2004.

[24] K. O. Keshishev, A. Y. Parshin, and A. V. Babkin. Experimental detection of crystallization waves in He-4. *JETP Lett.*, 30:56–59, 1979.

[25] S. Ulzega, A. Hofer, P. Moroshkin, and A. Weis. Measurement of the forbidden electric tensor polarizability of Cs atoms trapped in solid ^{4}He. *Phys. Rev. A*, 75:042505, 2007.

[26] P. L. Elliott, C. I. Pakes, L. Skrbek, and W. F. Vinen. Capillary-wave crystallography: Crystallization of two-dimensional sheets of He$^+$ ions. *Phys. Rev. B*, 61:1396–1409, 2000.

[27] J. Jortner, N. R. Kestner, S. A. Rice, and M. H. Cohen. Study of the properties of an excess electron in liquid helium. *J. Chem. Phys.*, 43:2614–2625, 1965.

[28] M. W. Cole and T. J. Sluckin. Nucleation of freezing by charged particles. I. Thermodynamics. *J. Chem. Phys.*, 67(2):746–750, 1977.

Chapter 4

Paper II: $6S_{1/2}$-$6P_{1/2}$ transition of Cs atoms in cubic and hexagonal solid ^{4}He

This paper presents calculational details of the extended bubble model performed entirely by myself. The model explains spectral shifts of the Cs atomic excitation and emission lines in liquid and solid He. The model calculations were also applied to interpret measurements of the Cs $6P_{1/2}$ state lifetime (Paper IV), and of the fine-structure and hyperfine-structure splittings.

My main contributions to the work were:

- Setting up of computer code using Mathematica 5.0 for the standard bubble model calculations. Extending the model to explain the sudden jumps of atomic excitation and emission lines at the liquid-solid phase transition. Including an additional spectral shift due to the interaction of the atomic dipole with its own radiation field (cavity effect).

- Performing measurement of the Cs D_1 emission line as a function of He pressure. Reanalyzing data of the hyperfine splitting. Calculating the hyperfine splitting using the bubble model. Calculating the fine structure splitting and comparison to experimental values.

- Producing figures, graphs and text for the paper.

$6S_{1/2}$-$6P_{1/2}$ transition of Cs atoms in cubic and hexagonal solid ^{4}He

A. Hofer[1], P. Moroshkin[1], S. Ulzega[2], D. Nettels[3], R. Müller-Siebert[4], and A. Weis[1]

[1] *Département de Physique, Université de Fribourg, Chemin du Musée 3, 1700 Fribourg, Switzerland*
[2] *EPFL, Lausanne, Switzerland.*
[3] *Biochemisches Institut, Universität Zürich, Switzerland*
[4] *SELFRAG AG, Langenthal, Switzerland*

Published in Phys. Rev. A **76**, 022502 (2007).

Abstract: We present a systematic experimental study of the absorption and fluorescence spectra of the $6S_{1/2} - 6P_{1/2}$ transition in Cs atoms isolated in solid ^{4}He matrices. The bubble model developed earlier for alkali atoms in liquid He is revised and applied to the present system. The analysis of the dependencies of absorption and fluorescence wavelengths on He pressure in liquid and solid He (cubic and hexagonal) environments leads us to modify the bubble model by taking the elastic deformation of solid He by the atomic bubble into account.

4.1 Introduction

Alkali atoms implanted in condensed He reside in nano-size spherical cavities - so-called atomic bubbles. These bubbles are formed around each impurity atom due to the Pauli principle that forbids any overlap between the closed S-shells of He atoms and the valence electron of the impurity. In the ground state of alkali atoms the valence electron is loosely bound in a spherically symmetric $nS_{1/2}$ orbital and the bubble is similar, albeit smaller, than that of a free electron in condensed helium. The spectroscopy of electrons in liquid [1, 2] and solid [3] He has been developed in the early 1990ies and the experimental results confirmed the predictions of the bubble model. Already in those studies it was found that the model originally developed for liquid He produces reliable results also for solid He - a consequence of the quantum nature of He crystals, where the He atoms are strongly delocalized. A similar tendency was observed in studies of absorption and emission spectra of alkali atoms (Cs and Rb) in liquid [4, 5] and solid [6] He, as well as for Ba atoms [7, 8]. However a more detailed analysis [9] of the spectral shift of the $6S_{1/2}$-$6P_{1/2}$ (D_1) transition of Cs atoms in He matrices as a function of He pressure reveals relatively large abrupt changes at the phase boundaries which can not be predicted by the bubble model.

A theoretical investigation of bubbles formed by Cs and Rb atoms in pressurized liquid He has been performed in [4, 5] and its results demonstrated a good agreement with experimental results. In the present paper we report on the results of systematic experimental investigations of the D_1 transition of Cs in solid He in a broad range of pressures, covering the body-centered cubic (bcc) and the hexagonal close-packed (hcp) crystalline phases. We also revise the bubble model and apply it to Cs atoms in solid He. We have included several effects not considered in the previous theoretical treatment of [4]: (i) the modification of the fine-structure splitting of Cs by the interaction with He; (ii) the interaction of the atomic dipole with its own radiation reflected at the bubble interface. We have further identified a contribution to the bubble energy due to elastic crystal deformations, which is not present in the case of liquid He, but which should be taken into account in solid He. The extended bubble model has allowed us to calculate absorption and emission spectra which are in good agreement with the experimental results obtained in liquid and solid He. We have applied our model calculations to the fine structure, the lifetime of the excited $6P_{1/2}$ state and the hyperfine splitting in the ground state of Cs in solid He and compare the results with experimental findings.

4.2 Theoretical model

4.2.1 Spherical bubble model

Our approach follows closely the one described in [4]. The essential feature of the standard bubble model (SBM) is the representation of the He matrix as a continuous medium, characterized by its density ρ and surface tension parameter σ. This treatment is justified not only for liquid, but also for solid He, which is a quantum crystal with a very large delocalization of the He atoms, and hence a strong overlap of their wavefunctions. The solid He matrix is so soft that the impurity atom imposes its own symmetry on the local trapping site. In particular, the spherically symmetric $6S_{1/2}$ and $6P_{1/2}$ electronic states of Cs in the cubic phase of solid He form bubbles of spherical shape.

Following the ideas of [10, 11] the many-body problem of the interaction of an alkali atom with a He atom can be reduced to a three body problem by assuming that the perturbations of the alkali core and the He atom are small compared to the perturbations of the alkali valence electron. We further use the fact that the alkali core and the He atom have closed shell structures. Details of these structures are not considered, but we assume that both can be polarized by the alkali valence electron. The properties of the atomic defect structure can then be described in terms of the mutual interactions of the valence electron, the "frozen" alkali core and the He atom(s). Using the Born-Oppenheimer approximation, the nucleus of the Cs and the He atom(s) can be treated as fixed in space and the problem is reduced to calculating the wavefunction of the valence electron in the combined potentials that it experiences. The geometry of the problem is sketched in Fig. 4.1.

The total potential felt by the alkali valence electron can be written as

$$V_{\text{tot}}(\mathbf{r}, \mathbf{R}) = V_{\text{Cs}}(\mathbf{r}) + V_{\text{He}}(\mathbf{r}, \mathbf{R}) + V_{\text{cross}}(\mathbf{r}, \mathbf{R}) + V_{\text{cc}}(\mathbf{R}),$$
(4.1)

where $V_{\text{Cs}}(\mathbf{r})$ and $V_{\text{He}}(\mathbf{r}, \mathbf{R})$ describe the interaction of the electron with the Cs core and the He atom, respectively. The cross term $V_{\text{cross}}(\mathbf{r}, \mathbf{R})$ describes the three body interaction, i.e., the polarization of the He atom by the Cs valence electron and the Cs$^+$ ion. Finally, the last term in Eq. 4.1, $V_{\text{cc}}(\mathbf{R})$, is the core-core interaction of the Cs$^+$ ion with the ground state He atom.

4.2 Theoretical model

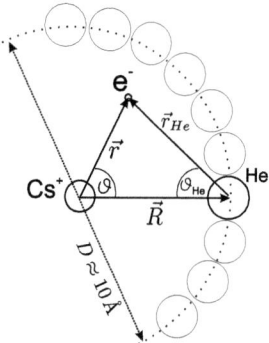

FIGURE 4.1: Schematic model of a Cs atom inside a spherical He bubble. The interaction is treated as a three body interaction between the Cs$^+$ core in the center of the bubble (origin of coordinate system), the valence electron of the Cs atom at position $\vec{r}$ and the He atom at position $\vec{R}$. The diameter of the bubble is approximately 1 nm for a ground state Cs atom.

4.2.2 Energy of the free Cs atom

The first term of Eq. 4.1 has the following contributions

$$V_{\mathrm{Cs}}(r) = V_{\mathrm{TF}}(r, \lambda) + V_{\mathrm{pol}}(r, r_c) + V_{\mathrm{so}}(r), \quad (4.2)$$

where $V_{\mathrm{TF}}(r, \lambda)$ is a scaled statistical Thomas-Fermi model potential. The core polarization due to the valence electron is represented by $V_{\mathrm{pol}}(r, r_c)$ and $V_{\mathrm{so}}(r)$ is the spin-orbit potential. For systems with many electrons an explicit calculation of the potential is not possible and one has to rely on approximative methods, of which the relativistic Hartree-Fock method has proven to be very successful. Hartree-Fock calculations are beyond our capabilities and we have opted for a simpler approach by using a scaled Thomas-Fermi model potential [12, 13] following the work of Norcross [14] to describe the interaction of the Cs valence electron with the Cs core. We have taken the Fermi-Amaldi correction for excluding the electrostatic self-energy of the electron and the exchange energy correction introduced by Dirac into account as described in [12]. This yields a corrected Thomas-Fermi potential $V_{\mathrm{TF}}(r, \lambda)$ with a scaling parameter λ which can be determined by fitting calculated level energies to the experimental level energies.

The core polarization potential with non-negligible dipole and quadrupole contributions can be written as [14]

$$\begin{aligned}V_{\mathrm{pol}}(r, r_c) &= -\frac{\alpha_d}{2r^4}\left[1 - e^{-(\frac{r}{r_c})^6}\right] \\ &\quad -\frac{\alpha_q - 3\beta_q}{2r^6}\left[1 - e^{-(\frac{r}{r_c})^{10}}\right].\end{aligned} \quad (4.3)$$

The values for the dipole, α_d, and quadrupole, α_q, core polarizabilities as well as for the dynamic correction β_q were taken from [14]. r_c represents a cutoff radius that depends on the angular momentum l of the valence electron and that is chosen together with the scaling parameter λ in order to match the experimental energies for the lowest lying states, i.e., the $6S_{1/2}$, $6P_{1/2}$, $6P_{3/2}$, $5D_{3/2}$ and the $5D_{5/2}$ state.

The spin-orbit potential $V_{\mathrm{so}}(r)$ is written with a relativistic correction as

$$V_{\mathrm{so}}(r) = \frac{\alpha^2}{4}\frac{1}{r}\frac{dV'(r)}{dr}\frac{1}{\left[1 + \frac{1}{4}\alpha^2 V'(r)\right]^2}\vec{L}\cdot\vec{S}. \quad (4.4)$$

The potential in Eq. 4.4 is $V'(r) = V_{\mathrm{TF}}(r, \lambda) + V_{\mathrm{pol}}(r, r_c)$ and α is the fine structure constant. The total potential $V_{\mathrm{Cs}}(r)$ seen by the free Cs atom's valence electron can then be used in the radial Schrödinger equation

$$-\frac{1}{2}\frac{d^2 u(r)}{dr} + \left[V_{\mathrm{Cs}}(r) + \frac{l(l+1)}{2r^2}\right]u(r) = E\,u(r) \quad (4.5)$$

to obtain the wavefunctions and eigenenergies of the free Cs atom. The total wavefunction of the valence electron is written as a product of radial and angular parts $\Psi(\mathbf{r}) = Y_{l,m}(\vartheta, \varphi)\,u(r)/r$. The boundary condition near the core is (see for example [15])

$$u(r) \propto r^{l+1} \text{ for } r \to 0. \quad (4.6)$$

For $r \to \infty$ we use the condition that the wavefunction has an exponential decay. All numerical calculations were performed with Mathematica 5.0. Figure 4.2 shows the calculated wavefunction for the ground and the first excited state of the free Cs atom. After having adjusted the parameters λ and r_c to yield the best agreement with experimental energies of the 5 lowest fine structure levels, the calculated energies of the states up to $n = 12$ were found to agree within 0.5% with their experimental values [16]. For higher lying states the values were compared to the values obtained using the hydrogen formula [17] with an effective principle quantum number n^* [16] and an agreement within 1% was obtained.

4.2.3 Cs-He interaction

The interaction of the Cs valence electron with the He atom, $V_{He}(\mathbf{r}, \mathbf{R})$, is treated in a similar way as its interaction with the Cs core in terms of a potential

$$V_{He}(\mathbf{r}, \mathbf{R}) = V_{e-He}(\mathbf{r}, \mathbf{R}) + V_{polHe}(\mathbf{r}, \mathbf{R}). \quad (4.7)$$

$V_{e-He}(\mathbf{r}, \mathbf{R})$ [18] is a pseudopotential that models the repulsion of the Cs valence electron when it enters the electronic cloud of the He atom due to the Pauli principle – the main reason for the bubble formation – and the incomplete screening of the nuclear charge of the He atom. This pseudopotential can be written as

$$V_{e-He}(\mathbf{r}, \mathbf{R}) = \sum_{l=0}^{\infty} \sum_{m=-l}^{l} V_l^{sr}(r_{He}) |Y_{lm}(\hat{r}_{He})\rangle\langle Y_{lm}(\hat{r}_{He})|, \quad (4.8)$$

in terms of a basis set of Gausian potentials $V_l^{sr}(r_{He}) = C_l \exp(-D_l r_{He}^2)$, centered on the He atom. The projection operators $|Y_{lm}(\hat{r}_{He})\rangle\langle Y_{lm}(\hat{r}_{He})|$ in Eq. 4.8 (l and m are the orbital momentum of the valence electron and its projection with respect to the He atom) are used to express the potentials with respect to coordinates centered on the Cs^+ core. The parameters C_l and D_l are taken from [18] and $\mathbf{r}_{He} = \mathbf{r} - \mathbf{R}$.

The second term of Eq. 4.7 can be written in analogy to the polarization potential of the free Cs atom in Eq. 4.3 as [18]

$$V_{polHe}(\mathbf{r}, \mathbf{R}) = -\frac{1}{2}\frac{\alpha_{dHe}}{(r_{He}^2 + r_{cHe}^2)^2}$$
$$-\frac{1}{2}\frac{\alpha_{qHe} - 6\beta_{qHe} + 2\alpha_{dHe}\, r_{cHe}^2}{(r_{He}^2 + r_{cHe}^2)^3}, \quad (4.9)$$

with the dipole polarizability α_{dHe}, the quadrupole polarizability α_{qHe} and a dynamic correction parameter β_{qHe}. $V_{polHe}(\mathbf{r}, \mathbf{R})$ shows the asymptotic r^{-4} and the r^{-6} dependences of the dipole and quadrupole polarizabilities and is screened at small distances by the parameter r_{cHe}.

A further contribution comes from the core-core interaction

$$V_{cc}(R) = V_{cc}^{rep}(R) - \frac{1}{2}\frac{\alpha_{dHe}}{(R^2 + r_{cHe}^2)^2}$$
$$-\frac{1}{2}\frac{\alpha_{qHe} - 6\beta_{qHe} + 2\alpha_{dHe}\, r_{cHe}^2}{(R^2 + r_{cHe}^2)^3}, \quad (4.10)$$

which describes the polarization of the He atom by the Cs^+ ion, and where $V_{cc}^{rep}(R) = a e^{-bR}$ is a repulsive potential, acting at small distances, where the electronic clouds of the two atoms start to overlap.

Finally we include a cross term $V_{cross}(\mathbf{r}, \mathbf{R})$ which represents the simultaneous polarization of the He atom by the Cs valence electron and the Cs^+ core

$$V_{cross}(\mathbf{r}, \mathbf{R}) =$$
$$f_{cutoff}(r, R)\left[-\frac{\alpha_{dHe}\cos\vartheta_{He}}{(R^2 + r_{cHe}^2)(r_{He}^2 + r_{cHe}^2)}\right.$$
$$\left.+\frac{1}{2}\frac{\alpha'_{qHe}\left(3\cos^2\vartheta_{He} - 1\right)}{(R^2 + r_{cHe}^2)^{3/2}(r_{He}^2 + r_{cHe}^2)^{3/2}}\right], \quad (4.11)$$

with the definitions $\alpha'_{qHe} = \alpha_{qHe} - 6\beta_{qHe} + 2\alpha_{dHe}\, r_{cHe}^2$, ϑ_{He} being the angle between $\mathbf{r}$ and $\mathbf{R}$. This term is needed to yield the correct behavior at large internuclear distances. The cutoff function $f_{cutoff}(r, R)$ is taken as

$$f_{cutoff}(r, R) = \begin{cases} 1 - e^{-(R/r-1)^2} & r \leq R \\ 0 & r > R \end{cases} \quad (4.12)$$

It assures that the cross term vanishes for small internuclear distances, where the electronic clouds overlap.

The parameter values used in the present calculations are taken from [18] and are listed in Table 4.1, together with our values for the parameters for the free Cs atom, which differ slightly from the ones used in [14].

4.2.4 Integration over the bubble and the bubble energy

We have now a complete expression that determines the interaction of the Cs valence electron with a single He atom. In order to calculate the interaction with all the helium atoms surrounding the Cs atom we treat the latter as an empty bubble in an incompressible fluid with a spherically symmetric density distribution $\rho(\mathbf{R})$

$$\rho(R, R_0, \epsilon)$$
$$= \begin{cases} 0 & R < R_0 \\ \rho_0[1 - \{1 + \epsilon(R - R_0)\}e^{-\epsilon(R-R_0)}] & R \geq R_0 \end{cases} \quad (4.13)$$

where R_0 is the bubble radius. ϵ describes the steepness of the density distribution at the bubble interface and ρ_0 is the bulk density $\rho(R \gg R_0)$ which depends on the He temperature and pressure.

4.2 Theoretical model

	Value (unit)	used in
α_{d}	19.03 (a_0^3)	$V_{\text{pol}}(r, r_c)$
α_{q}	118.26 (a_0^5)	$V_{\text{pol}}(r, r_c)$
β_{q}	19.18 (a_0^4)	$V_{\text{pol}}(r, r_c)$
r_c $(l=0)$	3.2272 (a_0)	$V_{\text{pol}}(r, r_c)$
r_c $(l=1)$	3.3.918 (a_0)	$V_{\text{pol}}(r, r_c)$
λ	1.07623	$V_{\text{TF}}(r, \lambda)$
α_{dHe}	1.3834 (a_0^3)	$V_{\text{polHe}}(\mathbf{r}, \mathbf{R})$ and $V_{\text{cc}}(R)$
α_{qHe}	2.3265 (a_0^5)	$V_{\text{polHe}}(\mathbf{r}, \mathbf{R})$ and $V_{\text{cc}}(R)$
β_{qHe}	0.706 (a_0^4)	$V_{\text{polHe}}(\mathbf{r}, \mathbf{R})$ and $V_{\text{cc}}(R)$
r_{cHe} $(l=0)$	1 (a_0)	$V_{\text{polHe}}(\mathbf{r}, \mathbf{R})$, $V_{\text{cc}}(R)$ and $V_{\text{cross}}(\mathbf{r}, \mathbf{R})$
a	49.1559	$V_{\text{cc}}^{\text{rep}}(R)$
b	1.8747 (a_0^{-1})	$V_{\text{cc}}^{\text{rep}}(R)$
C_l $(l=0)$	2.03	$V_1^{\text{sr}}(r_{\text{He}})$
D_l $(l=0)$	0.463 (a_0^{-2})	$V_1^{\text{sr}}(r_{\text{He}})$
C_l $(l \geq 0)$	-1	$V_1^{\text{sr}}(r_{\text{He}})$
D_l $(l \geq 0)$	1 (a_0^{-2})	$V_1^{\text{sr}}(r_{\text{He}})$

TABLE 4.1: Numerical values of parameters used for the numerical evaluation.

The energy needed to form a bubble is written in the commonly used way [8] as

$$E_{\text{bub}} = \frac{4}{3}\pi R_{\text{b}}^3 p + 4\pi R_{\text{b}}^2 \sigma + E_{\text{kin}}, \quad (4.14)$$

where p is the He pressure. The first term is the pressure volume work and the second term represents the energy of the surface tension. The third term is the volume kinetic energy due to the localization of the He atoms at the bubble interface. It is expressed as

$$E_{\text{kin}} = \frac{h}{16\pi\, m_{\text{He}}} \int d^3 R\, \frac{(\nabla \rho(R, R_0, \epsilon))^2}{\rho(R, R_0, \epsilon)}, \quad (4.15)$$

where m_{He} is the mass of the He atom. The radius R_{b} used in Eq. 4.14 is the center of mass of the bubble interface defined by the equation

$$\int_0^{R_{\text{b}}} \rho(R, R_0, \epsilon) R^2\, dR = \int_{R_{\text{b}}}^{\infty} [\rho_0 - \rho(R, R_0, \epsilon)] R^2\, dR. \quad (4.16)$$

In order to obtain electronic wavefunctions of the Cs atom confined in the bubble, we first integrate the potential over the He bulk

$$V_{\text{tot}}^{\text{bub}}(r, R_0, \epsilon) = V_{\text{Cs}}(r) + \int d^3 R\, \rho(R, R_0, \epsilon)\bigg[V_{\text{He}}(\mathbf{r}, \mathbf{R}) + V_{\text{cross}}(\mathbf{r}, \mathbf{R}) + V_{\text{cc}}(R)\bigg] \quad (4.17)$$

and then solve the radial Schrödinger equation (Eq. 4.5) by replacing V_{Cs} with $V_{\text{tot}}^{\text{bub}}$. The interaction of the Cs atom with an isolated He atom has no central symmetry. However, the integration over the bubble simplifies the problem since it leads to a central potential, so that the radial and angular variables can be separated.

The solutions depend on two parameters: R_0 and ϵ. In this way we do not only get the eigenenergy E_{int} depending on the bubble size but also the wavefunction of the valence electron $\Psi(r) = u(r)/r$. In Fig. 4.2 we compare the calculated radial wavefunctions for the $6S_{1/2}$ and the $6P_{1/2}$ states of the free Cs atom with those of a Cs atom in a spherical bubble. The wavefunctions of the atom in the bubble are slightly compressed by the bubble, the effect being more pronounced for the more extended $6P_{1/2}$ wavefunction.

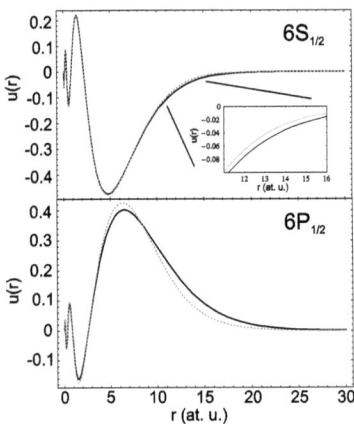

FIGURE 4.2: Calculated wavefunctions u(r) for the $6S_{1/2}$ and the $6P_{1/2}$ states of Cs for the free Cs atom (black **solid** curve) and for the Cs in the bubble (red **dashed** curve). The bubble parameters are $\epsilon = 2.45\,a_0^{-1}$ and $R_0 = 10.2\,a_0$, which correspond to the equilibrium bubble of the Cs ground state in bcc solid He. The inset in the upper graph shows the main difference between the two wavefunctions at a distance $13\,a_0$ from the nucleus.

The knowledge of the wavefunctions is important for the calculation of atomic properties like transition dipole moments, excited state lifetimes, or hyperfine structure. The equilibrium size of the bubble is determined via a numerical minimization of the total energy $E_{\rm tot}^{\rm bub}$ of the system by varying R_0 and ϵ. The total energy is the sum of the bubble energy (Eq. 4.14) and the interaction energy $E_{\rm int}$.

$$E_{\rm tot}^{\rm bub} = E_{\rm int} + E_{\rm bub}. \qquad (4.18)$$

The bubble parameters for both $6S_{1/2}$ and $6P_{1/2}$ states of Cs in bcc solid He close to liquid-solid phase boundary ($p = 26.9$ bar, $T = 1.5$ K) are compared with those calculated in [4] in Table 4.2. For the surface tension coefficient σ we use its measured value ($\sigma = 0.332$ dyne/cm) at $T = 1.5$ K at saturated vapor pressure [19]. No experimental data on its pressure dependence are available, and we assume that it is independent of pressure as discussed in the literature on electron bubbles [20], where different models for the pressure dependence of σ are suggested. In the range of parameters studied in our work the surface energy produces a contribution of about $\approx 10\,\%$ to $E_{\rm tot}^{\rm bub}$, so that our results are rather insensitive to variations of σ.

4.2.5 Hyperfine structure of Cs in solid He

As a first test of our model we have used the calculated $6S_{1/2}$ wavefunction to derive the bubble-induced change of the hyperfine coupling constant in the Cs ground state. The frequency of the corresponding hyperfine transition in bcc solid He has been measured earlier by our group [21]. It was found to be blue-shifted by approximately 196 MHz with respect to the free transition in free atoms (9192 MHz), with a slight pressure dependence.

The matrix elements of the hyperfine (Fermi contact) Hamiltonian $H_{\rm hf}$ in the cesium ground state are

$$H_{\rm hf} = A_{hf}\langle S, m_S, I, m_I|\mathbf{I}\cdot\mathbf{S}|S, m_S, I, m_I\rangle$$

with

$$A_{\rm hf} = -\frac{2\mu_0}{3}\frac{\mu_{\rm B}^2 g_I\, g_{\rm S}}{\hbar^2}\langle n,L,m_L|\delta(\mathbf{r})|n,L,m_L\rangle, \qquad (4.19)$$

The matrix element in Eq. 4.19 depends on the value of the wavefunction at the nucleus.

$$A_{hf} = -\frac{2\pi}{3}\alpha^2 g_I g_S|\Psi(0)|^2 \qquad (4.20)$$

and the hyperfine splitting of the ground state is $\delta\nu = 4\,A_{hf}$.

We first calculate the hyperfine splitting for the free Cs atom using the correction factors from [22] to account for relativistic effects and electrostatic and magnetic volume corrections. The obtained value ν=9770 MHz is approximately 6% larger than the experimental one. Our calculation of $\delta\nu$ for a Cs atom in a He bubble in the bcc phase shifts this value by 182 MHz to larger frequencies, which agrees within 6% with the experimentally measured shift of 196 MHz [21]. This increase of the hyperfine transition frequency is due to the compression of the electronic wavefunction by the surrounding pressurized He which increases $|\Psi(0)|^2$.

4.2.6 Fine structure of Cs in solid He

The $6P_{1/2}$ and $6P_{3/2}$ fine structure doublet in the free Cs atom is split by 554 cm^{-1}. In condensed helium this splitting cannot be studied in emission since the $6P_{3/2}$ state is quenched by the formation of exiplexes and a strong mixing with the $6P_{1/2}$

	R_0 (a_0)	ϵ (1/a_0)	R_b (a_0)	ΔE_S	ΔE_P	ΔE_{exc} (cm^{-1})	ΔE_{em} (cm^{-1})
$6S_{1/2}$ this work	10.22	2.45	11.06	186.8	780.6	593.8	149.5
$6P_{1/2}$ this work	12.95	1.99	13.99	501.7	651.2		
$6S_{1/2}$ from [4]	10.75	1.28	12.42	290.3	781.4	491	121
$6P_{1/2}$ from [4]	13.08	1.12	14.97	418.3	539.4		

TABLE 4.2: Comparison of our equilibrium bubble parameters with values from [4]. ΔE_{exc} and ΔE_{em} are the shifts of the excitation and emission lines respectively compared to the free atom. The values of this work are for a bcc crystal at T=1.5 K and p=26.9 bar. Values from [4] are for bcc at T=1.6 K and p=27.06 bar.

state [4, 9]. However, the transitions to both excited states can be studied via their absorption spectra.

The theoretical treatment of [4] neglects the effect of the He matrix on the fine structure splitting, although their experimental results show that the splitting increases with He pressure, reaching $\Delta = 670$ cm^{-1} at 20 bar. Our experimental results show a further increase up to $\Delta = 700$ cm^{-1} in hcp solid He at 30 bar [9]. The theoretical model presented above allows us to calculate this splitting as the difference between the eigenenergies of the perturbed $6P_{1/2}$ and $6P_{3/2}$ states in a spherical bubble formed around the ground state Cs atom. For liquid He at 25 bar we obtain $\Delta = 642$ cm^{-1} and for bcc solid $\Delta = 649$ cm^{-1}. In both cases the splitting is underestimated, however the sign of the shift and its order of magnitude are predicted correctly. The theoretical model used in this work is refined with respect to the one of [4] as it takes the bubble effect on the fine structure splitting into account.

4.2.7 Lifetime of the $6P_{1/2}$ state

With the theoretical model presented above we also calculate the lifetime τ of the excited $6P_{1/2}$ state of Cs in condensed He. Experimental data on the dependence of τ on He pressure are available for superfluid He [23] up to the solidification point. Recently we have measured lifetimes in bcc and hcp solid He up to $p = 36$ bar [24]. The results of [23, 24] show that in liquid and bcc solid He, τ has a pressure independent value of 32.5 ns, 2.3 ns shorter than the lifetime in a free Cs atom. At the phase transition to the hcp phase the lifetime shortens by 3.2 ns and further decreases with increasing He pressure.

The radiative lifetime τ of an excited state is related to the transition dipole moment $|\langle 6P_{1/2}\|er\|6S_{1/2}\rangle|$ and frequency ω_0 via

$$\frac{1}{\tau} = \frac{\omega_0^3 e^2}{3\pi\epsilon_0 \hbar c^3} \frac{1}{2} |\langle 6P_{1/2}\|r\|6S_{1/2}\rangle|^2. \quad (4.21)$$

We have calculated the transition dipole moment using the wavefunctions of the $6P_{1/2}$ and $6S_{1/2}$ states of Cs perturbed by the bubble, as discussed in detail in [24]. The results show that the dipole moment decreases with increasing pressure. However, this change is largely compensated by a simultaneous increase of the transition frequency (the blueshift discussed in the following subsection) and the resulting lifetime is almost constant in agreement with the experimental data in liquid and bcc solid He. We have also shown [24] that the reduction of τ with respect to its free atomic value is due to the interaction of the atomic dipole with its own radiation field reflected at the bubble interface (cavity effect). In the case of hcp matrices the observed pressure dependence of τ is attributed to the onset of a pressure-dependent radiationless formation of exciplex [24].

4.2.8 The cavity effect

The above treatment has not yet taken into account that the excited Cs atom interacts with its own electromagnetic radiation reflected at the bubble interface. It is well known that a static (or oscillating) electric dipole close to a dielectric interface induces a static (or oscillating) polarization in the dielectric. The interaction between the dipole and its mirror image in the dielectric results in a redshift of the emitted light, and affects the lifetime of the atomic oscillator [24]. The problem of an excited atom interacting with a spherical microcavity in a dielectric has been treated in [25]. The shift of the transition frequency is given by

$$\delta = -\frac{e^2}{2\omega_0 m_0 d^2} Re[\mathbf{d}\cdot\mathbf{E}], \quad (4.22)$$

where m_0 is the electron mass, ω_0 the transition frequency, $\mathbf{d}$ the transition dipole moment, and $\mathbf{E}$ the field produced by the polarized dielectric at the position of the atom. We have evaluated this expression and calculated the corresponding correction to the transition frequency. For a given bubble configuration we made a numerical evaluation of the induced polarization in the surrounding solid He and calculated the field produced by that polarization at the center of the bubble. Retardation effects can be neglected because of the small bubble size. For bcc solid He at 1.6 K we obtain a redshift of 44 cm^{-1} for the emission line and of 83 cm^{-1} for the absorption line. This cavity effect is taken into account in the calculated lineshifts presented in Fig. 4.5. The same approach was used in [26] to calculate wavelength corrections of the absorption line of alkali atoms (Li, Na, K) bound in a dimple at the surface of He nano-droplets. In that case the correction was much smaller, about 9 cm^{-1}, due to the loosely bound structure of the trapping site.

4.2.9 Line shape of absorption and emission lines

A standard way [27] for calculating the lineshape of the $6S_{1/2}$-$6P_{1/2}$ absorption line consists in considering the smearing out of the ground state wavefunction due to bubble oscillations. Here we consider only radial (breathing mode) oscillations around the equilibrium bubble radius $R_0(6S)$, whose wave function can be obtained in the following way. Figure 4.3 shows the total bubble energy E_{tot}^{bub} as a function of the bubble radius. We use this energy (and not just its harmonic approximation near the minimum) as the potential in a one-dimensional Schrödinger equation. The solutions then yield the eigenenergies and wavefunctions of the oscillations. We consider the mass of the oscillator to be the hydrodynamic mass of the bubble $M_{eff} = 4\pi R_b^3 \rho_0 m_{He}$.

The splitting between the vibrational ground state and the first excited vibrational level is equivalent to 7.5 K at 26.6 bar, so that at the temperature $T = 1.6$ K of the experiment, only the lowest vibrational state is populated. The probability distribution for finding a bubble with radius R_0 is then given by $|\phi_0(R_0)|^2$, where $\phi_0(R_0)$ is the wavefunction associated with the corresponding zero-point energy, whose R_0 dependence is shown in Fig. 4.3. To each bubble radius R_0 corresponds a given transition energy with a relative weight given by $|\phi_0(R_0)|^2$. An equivalent procedure can be applied for calculating the emission spectra. In that case one starts from breathing mode oscillations of the bubble around $R_0(6P)$.

The shape and the size of the bubble do not change during the electronic transition since the transition occurs on a time scale shorter than the bubble oscillation period (Frank-Condon principle). Once the Cs atom is excited the bubble relaxes to a larger radius that reflects the larger extension of the excited state wavefunction. One can estimate that this relaxation occurs on a picosecond time scale. The fluorescence transition occurs in the larger bubble in which the excited state lives for a few ten ns, close to the free atomic lifetime.

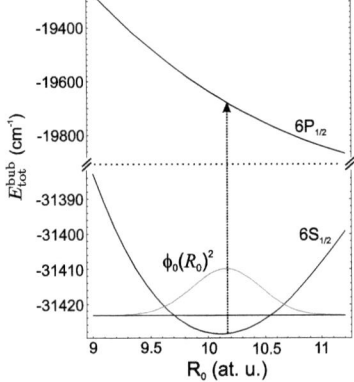

FIGURE 4.3: Calculated potential energy (solid black line) of the Cs $6S_{1/2}$ ground state in the spherical bubble including the bubble energy as a function of the bubble radius R_0 with the probability distribution $\phi_0(R_0)^2$ shown as red curve.

In Fig. 4.4 we compare the theoretical excitation and emission lineshapes for the D_1 ($6S_{1/2}$ - $6P_{1/2}$) line in a bcc He crystal to the experimentally measured spectra. Both absorption and emission lines are blue shifted and broadened with respect to the free atomic line. The repulsive interaction between the valence electron and the bubble interface shifts both atomic states towards higher energies. Since the electronic wavefunction of the excited state has a larger radial extension the shift of that state is larger. As a result the net transition energy increases and the lines become blue shifted. The shift and broadening are more pronounced in absorption since it occurs in a bubble of smaller size than the bubble in which the emission occurs. The calculated transition wavelengths, defined as the numerically evaluated centers of gravity of the lines are shown as a function of He pressure in Fig. 4.5 together with experimental results for pressures ranging from liquid He (HeII), via the crystalline bcc phase to the hcp phase. As a general trend the transition frequencies increase with increasing He pressure due to the increased perturbation of the Cs atom by the He matrix.

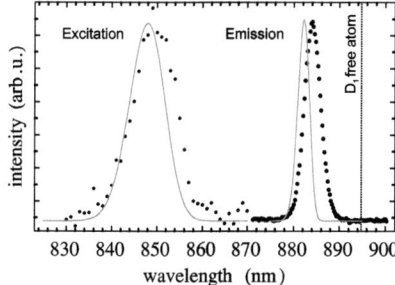

FIGURE 4.4: Experimental (black dots) and theoretical (solid red line) excitation and emission spectrum of the $6S_{1/2}$ - $6P_{1/2}$ transition in the bcc phase of solid He. Experimental conditions: $T \simeq 1.6$ K, $p = 27.8$ bar. Note the good agreement of the line positions within the experimental line widths. The line positions as a function of He pressure are shown in Fig. 4.5.

4.3 Experiment

The experiments were performed in a Cs doped solid He matrix. Details of the technique for doping a He crystal were presented in previous publications [28, 29]. A He crystal is produced by pressurizing liquid He in a copper cell immersed in a liquid He bath cooled to 1.6 K by pumping on its surface. The cell has five windows in three orthogonal directions for optical access. The crystal is doped with Cs atoms by laser ablation using a frequency doubled pulsed Nd:YAG laser (532 nm, repetition rate ∼3 Hz, pulse energy 10 mJ). The laser beam is focused by a height adjustable lens mounted above the cell onto a solid Cs target located at the bottom of the cell. A cw diode laser at 850 nm, or the idler output of a tunable optical parametric oscillator (OPO) pumped by the third harmonic of a Nd:YAG laser were used for the optical excitation of the implanted atoms. The idler beams of the OPO can be tuned over the range 770–1100 nm. The atomic fluorescence light is collected by a lens inside the cryostat and collimated into a direction perpendicular to the ablation and excitation laser beams, where it is focused into a grating spectrograph (MS257, Oriel) equipped with a CCD camera.

All measurements were done at 1.5 K or 1.6 K, in the pressure range of 26 - 38 bar, either in the bcc or hcp phase of solid He.

Excitation spectra. The excitation spectrum of the D_1 transition recorded in the bcc phase at 27 bar is shown in Fig. 4.4. This spectrum was obtained by tuning the OPO over the range of 820 - 870 nm in 1 nm steps, while recording the emitted fluorescence near 880 nm. The measurements were repeated for different helium pressures. At each pressure the center of gravity of the excitation band was determined numerically. The dependence of the line centers λ_{ex} as a function of He pressure are shown in Fig. 4.5(a), where they are compared to the theoretical predictions.

The experimental pressure dependence of the absorption (and emission) lines presented in Fig. 4.5 can be compared to corresponding measurements in pressurized superfluid He [4] shown in the same plot. In HeII the absorption line shows an almost linear shift towards shorter wavelengths with increasing He pressure, and the data in solid He have a pressure shift with a practically identical slope. This common slope is very well predicted by the SBM calculations made in this work (solid lines in Fig. 4.5).

A prominent feature in Fig. 4.5 is the large jump δ_{ex} of the excitation wavelength at the boundary between the liquid and solid phases. At the liquid-bcc phase transition the He density ρ_0 increases by about 8%. In the bubble model this change of density yields a blueshift of the excitation line by approximately 2 nm (28 cm^{-1}), much smaller than experimentally observed jump of 10 nm (140 cm^{-1}). The excitation line shows another jump – of opposite sign – at the bcc–hcp phase transition. At this point the helium density increases by 0.4%, for which the SBM predicts a blueshift of 0.2 nm (3 cm^{-1}), not visible on the scale of Fig. 4.5, whereas the measured λ_{ex} shifts by 1.5 nm (20 cm^{-1}) to the red. We interpret the sign and magnitude of this jump as being due to the static quadrupolar deformation of the atomic bubble in the uniaxial hcp crystal [29]. Clearly, the bubble model assuming spherical bubble shapes is not capable of treating this phenomenon and the development of an extended bubble model that takes bubble deformations into account is in progress.

Emission spectra. We have also measured the pressure dependence of the D_1 emission line center λ_{em} excited at a fixed wavelength (850 nm). The excitation wavelength was not adjusted when changing the pressure since the absorption line is rather broad (Fig. 4.4). The results are shown in

Fig. 4.5(b), where the positions of the emission lines are taken as their centers of gravity.

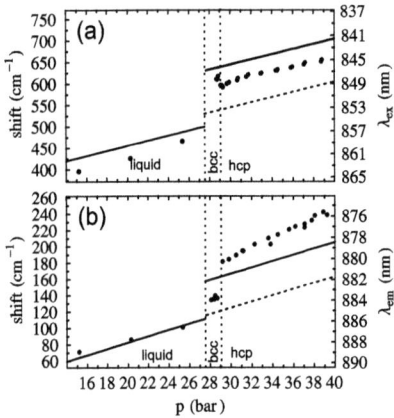

FIGURE 4.5: Experimental (dots) excitation (a) and emission (b) wavelengths of the $6S_{1/2} \to 6P_{1/2}$ (D_1) transition. Experimental data in liquid He are taken from [4]. The solid lines are calculated using the extended SBM presented in this work (the extension of the model is discussed in Sect. 4.4.1) including the cavity shift. The dashed lines show the predicted jumps at the liquid-solid phase transition without using the extended model (only SBM). Vertical dashed lines mark the phase transitions of condensed He.

The pressure shift of the emission line λ_{em} (Fig. 4.5(b)) also shows a linear dependence on He pressure. However, the slope is now different in the liquid and solid phases. The spherical bubble model gives a very good agreement with the slope and the absolute values in liquid He and slightly underestimates the one in solid He. As in the case of the absorption line, the blue jump $\delta_{em} = 2$ nm (30 cm^{-1}) observed at the HeII-bcc transition can not be explained by the spherical bubble model as being due to the 8% increase of density. Another remarkable fact is that the jump of the emission line at the bcc-hcp phase transition is towards shorter wavelengths. The SBM predicts the same sign of the jump, but strongly underestimates its magnitude.

4.4 Discussion

4.4.1 Extension of the bubble model

The calculations using the spherical bubble model presented here as well as those reported in [4] can not reproduce the large jump of the excitation and emission lines at the liquid-solid phase transition observed in the experiments. In this subsection we suggest an extension of the SBM which explains this experimental observation.

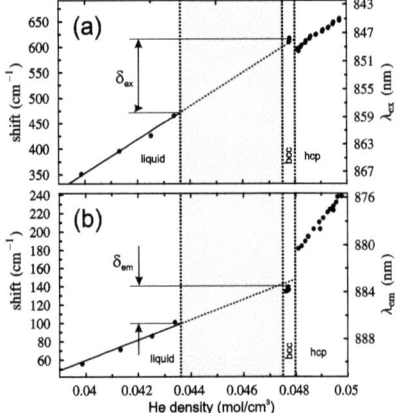

FIGURE 4.6: Experimental excitation (a) and emission (b) wavelengths(dots) of the D_1 line as a function of the He density. Data in liquid He are taken from [4]. The solid lines are fits to the data in the liquid phase. Vertical dotted lines mark phase boundaries. The grey region represents the experimentally non-accessible densities corresponding to the density jump at the liquid-bcc transition.

In Fig. 4.6 λ_{ex} and λ_{em} (the same data as Fig. 4.5) are plotted as functions of the He density ρ rather than of the He pressure. Both density dependencies are linear since in liquid and solid He the density is nearly proportional to the pressure (except for the discontinuity at the phase boundary). The striking feature of Fig. 4.6 is that the data points measured in the bcc phase and those in liquid He lie exactly on the same straight line, without any jump at the phase boundary. This observation suggests that it is the density rather than the pressure, which is responsible for the "pressure" shift. At the same time, the shift at the bcc-hcp phase boundary has a different nature and is most likely related to the change in the bubble shape due to the anisotropy of the crystal.

In liquid He the equilibrium bubble radius R_0 is established as a balance between the repulsive Cs-He interaction (discussed in section 4.2.3) and the bubble energy E_{bub}, which is minimized for $R_0 \to 0$ (Eq. 4.14). The former is proportional to the He density ρ (see eq. 4.17), while the latter is dominated by the pV term and is thus proportional to the He pressure. An increase of the He pressure is accompanied by a corresponding increase of the He density and both E_{bub} and E_{int} increase. Therefore the total energy increases with a very small change (decrease) of the bubble radius. At the phase boundary the density increases by 8% without any change in the pressure. According to the SBM a new equilibrium bubble with a larger value of R_0 is established with a relatively small corresponding increase in energy. The undisturbed linear dependence of Fig. 4.6 suggests that the bubble stays frozen or even shrinks at the phase transition due to an additional force which compensates the increased Cs-He repulsive force.

We identify this new force as an elastic restoring force which appears in the solid compressed by the expansion of the bubble at *constant volume* of the sample. It is not present in liquid He, where the formation of the bubble proceeds by a displacement of helium at *constant pressure*. In order to calculate this restoring force we use the fact that in equilibrium it compensates the force acting on the bubble interface from inside.

$$F_{\text{elastic}} = -F_{\text{Cs-He}} = -\left.\frac{\partial E_{\text{pot}}}{\partial R_0}\right|_{R_0=R_{\text{eq}}} \quad (4.23)$$

where R_{eq} is the equilibrium bubble radius for a given electronic state (here $6S_{1/2}$ or $6P_{1/2}$). The energy of the deformation is then given by $F_{\text{elastic}}\Delta R$, where $\Delta R = R_0^{\text{solid}} - R_0^{\text{liquid}}$ is the difference between the equilibrium bubble size in solid and in liquid He. Equation 4.14 defining the bubble energy becomes

$$E_{\text{bub}} = \frac{4}{3}\pi R_b^3 p + 4\pi R_b^2 \sigma + E_{\text{kin}} - \left.\frac{\partial E_{\text{pot}}}{\partial R_0}\right|_{R_0=R_0^{\text{liquid}}} \Delta R. \quad (4.24)$$

The additional term leads to an equilibrium bubble size in bcc He which is slightly smaller than the bubble size in liquid. We have recalculated the spectra using the extended expression (4.24) for E_{bub}. The resulting pressure dependencies of the excitation and emission wavelengths are shown in Fig. 4.5 by solid lines. In both cases we obtain a much better agreement with the experimental results in He II and in bcc solid He than with the standard spherical bubble model (shown by dashed lines in Fig. 4.5).

The jump of excitation and emission lines at the bcc-hcp phase transition remains when plotted against density and cannot be explained, even in the frame of the present extension of the standard bubble model. The fundamental difference between the bcc and the hcp phases is the crystalline symmetry: the uniaxial hexagonal phase has strongly anisotropic elastic constants which affect the shape of the atomic bubble. In consequence the potential seen by the valence electron is no longer a central potential. Some aspects of static anisotropic bubble deformations are discussed in [29]. A more detailed calculation of this effect on atomic spectra is currently underway and will be presented in a forthcoming publication.

4.4.2 Summary

We have presented a detailed discussion of the spherical bubble model for Cs atoms in liquid and solid ^{4}He with an important extension of the model that allows us to explain the sudden jumps of absorption and emission wavelenghts of the Cs $6S_{1/2} - 6P_{1/2}$ transition at the liquid-solid phase transition. The extension of the model includes an additional bubble energy term for the solid phase that accounts for the elastic properties of the crystal and we included an additional shift due to the interaction of the excited Cs atom with its own radiation field (cavity effect). The extended bubble model gives an excellent agreement within the experimental linewidths of the absorption and emission wavelengths of Cs atoms in liquid and solid bcc He. We have shown that it is the density of the He matrix rather than its pressure that is responsible for the line shifts. The model calculations were also applied to measurements of the lifetime of the Cs $6P_{1/2}$ state [24], the fine structure and the hyperfine splitting in the Cs ground state and gave good agreement with experimental results. A model for the deformations of the atomic bubbles in the uniaxial hcp phase of solid He is in progress and will be the subject of a future publication.

ACKNOWLEDGMENTS This work was supported by the grant number 200021-111941/1 of the Schweizerischer Nationalfonds.

References

[1] C. C. Grimes and G. Adams, Phys. Rev. B **45**, 2305 (1992).

[2] A. Y. Parshin and S. V. Pereverzev, JETP **74**, 68 (1992).

[3] A. I. Golov and L. P. Mezhov-Deglin, J. Exp. Theor. Phys. Lett. **56**, 514 (1992).
[4] T. Kinoshita, K. Fukuda, Y. Takahashi, and T. Yabuzaki, Phys. Rev. A **52**, 2707 (1995).
[5] T. Kinoshita, K. Fukuda, and T. Yabuzaki, Phys. Rev. B **54**, 6600 (1996).
[6] S. Kanorsky, A. Weis, M. Arndt, R. Dziewior, and T. Hänsch, Z. Phys. B **98**, 371 (1995).
[7] S. I. Kanorsky, M. Arndt, R. Dziewior, A. Weis, and T. W. Hänsch, Phys. Rev. B **49**, 3645 (1994).
[8] S. I. Kanorsky, M. Arndt, R. Dziewior, A. Weis, and T. W. Hänsch, Phys. Rev. B **50**, 6296 (1994).
[9] P. Moroshkin, A. Hofer, D. Nettels, S. Ulzega, and A. Weis, J. Chem. Phys. **124**, 024511 (2006).
[10] J. Pascale and J. Vandeplanque, J. Chem. Phys. **60**, 2278 (1974).
[11] G. Peach, J. Phys. B **11**, 2107 (1978).
[12] P. Gombas, *Pseudopotentiale* (Springer-Verlag Berlin-Gottingen-Heidelberg, 1956).
[13] B. H. Bransden and C. J. Joachian, *Physics of Atoms and Molecules* (Longman, London, New York, 1983).
[14] D. W. Norcross, Phys. Rev. A **7**, 606 (1973).
[15] C. Cohen-Tannoudji, *Quantum Mechanics*, vol. 1 (Hermann and, 1977).
[16] K. H. Weber and C. J. Sansonetti, Phys. Rev. A **35**, 4650 (1987).
[17] I. I. Sobelman, *Atomic spectra and radiative transitions*, Springer series on atoms and plasmas (Springer, 1992).
[18] J. Pascale, Phys. Rev. A **28**, 632 (1983).
[19] M. Iino, M. Suzuki, and A. J. Ikushima, J. Low Temp. Phys. **61**, 155 (1985).
[20] M. Rosenblit and J. Jortner, J. Phys. Chem. A **101**, 751 (1997).
[21] S. Lang, S. Kanorski, M. Arndt, S. B. Ross, T. W. Hänsch, and A. Weis, Europhys. Lett. **30**, 233 (1995).
[22] H. Kopfermann, *Kernmomente* (Akademische Verlagsgemeinschaft G.M.H, Frankfurt am Main, 1956).
[23] T. Kinoshita, K. Fukuda, T. Matsuura, and T. Yabuzaki, Phys. Rev. A **53**, 4054 (1996).
[24] A. Hofer, P. Moroshkin, S. Ulzega, and A. Weis.
[25] V. V. Klimov, M. Ducloy, and V. S. Letokhov, J. Mod. Opt. **43**, 549 (1996).
[26] F. Stienkemeier, J. Higgins, C. Callegari, S. I. Kanorsky, W. E. Ernst, and G. Scoles, Z. Phys. D **38**, 253 (1996).
[27] S. Kanorsky and A. Weis, *Atoms in nanocavities*, vol. 314 (Kluwer Academic Publisher, 1995).
[28] M. Arndt, R. Dziewior, S. Kanorsky, A. Weis, and T. Hänsch, Z. Phys. B **98**, 377 (1995).
[29] P. Moroshkin, A. Hofer, S. Ulzega, and A. Weis, Fiz. Nizk. Temp. **32**, 1297 (2006), (Low Temp. Phys., 32, 981 (2006)).

Chapter 5

Paper III:
Lifetime of the Cs $6P_{1/2}$ state in bcc and hcp solid ^{4}He

This paper presents the first measurement of pressure and crystal structure dependent lifetimes of the Cs $6P_{1/2}$ state in solid ^{4}He compared to predictions by the spherical bubble model.

My main contributions to the work were:

- Setting up the optics and electronics for the corelated photon counting measurements. This includes the installation and testing of the photomultiplier and the installation and calibration of the different electronic components for the correlated measurements. The output of the time-to-amplitude converter (TAC) is read out by a multi-scaler card in a PC. We use a commercial spectroscopic software for the control of that card.
- Recording the data together with my colleague P. Moroshkin.
- Analyzing the experimental data with Mathematica.
- Developing the theoretical model (extended bubble model) that yields the theoretical predictions for the Cs $6P_{1/2}$ state lifetime in bcc solid ^{4}He.
- Producing figures, graphs and text for the paper.

Lifetime of the Cs $6P_{1/2}$ state in bcc and hcp solid ^{4}He

A. Hofer[1], P. Moroshkin[1], S. Ulzega[2] and A. Weis[1]

[1]*Département de Physique, Université de Fribourg, Chemin du Musée 3, 1700 Fribourg, Switzerland*
[2]*EPFL, Lausanne, Switzerland.*

Published in Europhys. J. D (2007), DOI: 10.1140/epjd/e2007-00275-5.

Abstract: We present the first experimental study of time-resolved fluorescence from laser-excited Cs($6P_{1/2}$) atoms isolated in a solid ^{4}He matrix. The results are compared to the predictions of the bubble model including the interaction of the atomic dipole with its radiation reflected at the bubble interface. Our results show that in liquid He as well as in the body-centered cubic (bcc) crystalline phase of He the lifetime of excited Cs atoms does not depend on He pressure, in agreement with our theory. When going from the bcc to the hexagonally close-packed (hcp) phase of ^{4}He the lifetime is reduced by $\approx$10% and decreases further whith increasing He pressure. We assign this effect to the formation of Cs*He$_n$ exciplexes, and determine the pressure dependence of the probability that the $6P_{1/2}$ state decays via this nonradiative channel.

5.1 Introduction

In recent years the study of metal atoms, and in particular alkali atoms, isolated in condensed ^{4}He matrices [1, 2, 3, 4] and in He nanodroplets [5] has become a well established spectroscopic technique. In the past most of the information on the dopants' properties were inferred from spectroscopic experiments involving the study of absorption and emission spectra of laser-induced fluorescence from the near UV to the near IR range of the spectrum as well as from double resonance experiments involving optically detected magnetic resonance induced by rf or microwave radiation. Overviews of the fields are given in the review papers [6] (liquid helium and nano-droplets) and [7] (solid helium). Recent aspects of the dopant spectroscopy in condensed helium and He droplets include the observation of bound states of excited alkali atoms with one or several bound He atoms, so-called exciplexes [7].

No time dependent studies of alkalis in *solid* He have been reported so far, although excited state lifetimes of Rb and Cs immersed in *superfluid* He have been studied in the past [8, 9]. Those experiments had revealed some interesting observations: while the Cs $6P_{1/2}$ lifetime was found to be pressure independent up to the solidification point, with a value below the free atomic lifetime, the lifetime of the corresponding $5P_{1/2}$ state of Rb decreasd with increasing pressure, leading to a complete quenching of the fluorescence. The authors of [9] interpreted their observations as being due to the formation of exciplexes, a deexcitation channel which opened only for Rb as the corresponding potential barrier for exciplex formation is lower than for Cs.

In this paper we present a first study of pressure dependent lifetimes of the $6P_{1/2}$ state in Cs atoms implanted in the body-centered cubic (bcc) and in the hexagonally close-packed (hcp) phase of solid He. In the bcc phase we find lifetimes which coincide with the ones observed in the liquid phase, while the passage to the hcp phase shows a pronounced jump to lower lifetimes, with a further quenching when the He pressure is increased.

In contrast to He droplets, in which the alkali atoms reside on the droplet surface and quickly desorb after optical excitation, the alkali atoms, when immersed into bulk (liquid or solid) He matrices, are stabilized in small cavities, so-called atomic bubbles. These bubble structures (spherical in bcc, slightly deformed in hcp) can be modelled by the so-called standard bubble model (SBM).

In Section 5.2 we will recall the main aspects of

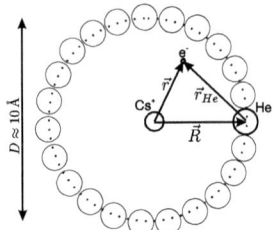

FIGURE 5.1: Ingredients for the calculation of the Cs wavefunctions in a spherical He bubble. The Schrödinger equation is solved for the Cs valence electron moving in the combined potential of the Cs$^+$ core and the pseudopotential of the surrounding He bulk. The size of the bubble is approximately 10 Å for a ground state Cs atom.

the standard bubble model (SBM) used to calculate excited state lifetimes. In Section 5.3 we present the experimental setup and the lifetime measurements and the results of the measurements are discussed in Section 5.4.

5.2 Theory

We have applied the standard bubble model (SBM) [1, 10] to describe the bubble structure and the properties of the trapped atom. The SBM treats the He bulk as an incompressible liquid and the bubble shape is described by a spherical model distribution of the He density. The size of the bubble for a given pressure is obtained by minimizing the total bubble energy whose standard [10] contributions are the pressure-volume energy, the surface tension energy, a kinetic energy contribution related to the density change at the interface and the Cs-He interaction energy.

The latter contribution can be described by a summation of Cs-He pair potentials. Here, however, we take another approach, by solving the Schrödinger equation for the valence electron moving in the potential of the Cs$^+$ core and a pseudopotential formed by the surrounding He (Fig. 5.1) [11, 12]. A coarse outline of this calculation was given in [13] and will be presented with more details in [14]. The main contribution to the interaction of the electron with the core is described by a scaled Thomas-Fermi model potential following the work of Gombas [12] and Norcross [15]. Several corrections, such as the dipolar and quadrupolar core polarizabilities were taken into account.

The radiative lifetime of an excited state with angular momentum J_e and transition frequency ω_0 to the ground state (angular momentum J_g) is related to the transition dipole matrix element via

$$\frac{1}{\tau} = \frac{\omega_0^3}{3\pi\epsilon_0\hbar c^3} \frac{1}{2J_e+1} |\langle n_e L_e J_e \| d \| n_g L_g J_g \rangle|^2, \quad (5.1)$$

where the reduced matrix element is related to the radial integral of the wavefunctions via

$$|\langle n_e L_e J_e \| d \| n_g L_g J_g \rangle| = |\langle 6P_{1/2} \| d \| 6S_{1/2} \rangle|$$
$$= \sqrt{\frac{2}{3}}\, e \int_0^\infty R_{6S}\, r^3\, R_{6P}\, dr\,. \quad (5.2)$$

We evaluated the radial integrals numerically using our solutions of the Schrödinger equation. The obtained values for the free Cs atom and for Cs in a spherical He bubble are listed in Table 5.1, together with the transition wavelengths and the lifetimes. For the free Cs atom we obtain a lifetime which is approximately 4% larger than the precise experimental value of [16, 17]. This sets a scale for the precision of our model calculations.

We have also calculated the lifetime of the $6P_{1/2}$ state in spherical He bubbles for pressures covering the superfluid and the solid bcc phase. According to the bubble model the optical excitation to the $6P_{1/2}$ state occurs in a (small) bubble whose size is determined by the $6S_{1/2}$ wave function. Once the atom is in the excited state the bubble expands in order to minimize the total energy due to the larger spatial extension of the $6P_{1/2}$ state. This expansion happens on a much faster time scale than the subsequent fluorescence emission, which occurs in the larger bubble, in which the Cs wavefunctions are less perturbed by the He matrix. The calculated lifetimes τ^{theo} are shown, together with the experimental data, in Fig. 5.4.

As a general trend we observe that the dipole matrix element decreases with increasing He pressure, while the transition frequency ω_0 increases, so that the two changes cancel each other to a large extend. This reflects well the observed independence of the $6P_{1/2}$ lifetime in pressurized superfluid He [9] and its present extension into the solid bcc phase discussed below.

For a complete description one has to take the interaction of the excited Cs atom with its own electromagnetic radiation reflected at the interface to the (dielectric) helium bulk into account. It is well known that an oscillating electric dipole close to a dielectric surface induces an oscillating polarization in the dielectric. The interaction between the two dipoles then results in a frequency shift of the emitted light and in a change of its lifetime (real and imaginary parts of the dipole's energy). The latter effect depends on the retardation between the oscillations of the atomic dipole and the reflected electromagnetic field. It may vary considerably and even change sign depending on the thickness of the bubble interface and its refraction index [18]. In order to estimate the influence of the reflected radiation on the observed lifetime of Cs in condensed He we use the results of [18], where an analytic expression is derived for the radiative lifetime of an atom in a spherical nanocavity with a sharp boundary to the outer dielectric bulk. Assuming an index of refraction for solid He of $n_{\text{He}} = 1.0365$ (bcc phase at 1.5 K, 26.6 bar, molar volume 21.1 cm^3) and a bubble radius $R_b(6P_{1/2}) = 14.0\,a_0$, obtained from our bubble calculations [14], we find an enhanced radiative decay rate $\gamma = 1.086\,\gamma_0$, where γ_0 is the uncorrected radiative decay rate of the Cs atom. The lifetime corrected for this cavity effect is given in Table 5.1.

Note that this dielectric cavity correction was shown previously to contribute to the shift of spectral lines of alkali atoms on He nano-droplets [5].

5.3 Experiment

5.3.1 Experimental setup

The experiments were performed in a Cs doped solid He matrix. Details of the techniques for doping a He crystal were presented in our previous publications [19, 7]. A crystal is produced by pressurizing liquid He in a copper cell (170 cm^3 of inner volume) immersed in a liquid He bath cooled to 1.5 K by pumping on its surface. The cell has five windows in three orthogonal directions for optical access. The crystal is doped with Cs atoms by laser ablation using a frequency doubled pulsed Nd:YAG laser (10 mJ pulses) with a repetition rate of ∼3 Hz. The laser beam is focused by a height adjustable lens mounted above the cell onto a solid Cs target located at the bottom of the cell.

The setup for the lifetime measurements of the atomic $6P_{1/2}$ state of Cs in solid He is shown in Fig. 5.2. A pulsed diode laser at 850 nm (10 kHz repetition rate) was used for the optical excitation of the implanted atoms. The atomic fluorescence light is collected by a lens inside the cryostat and collimated into a direction perpendicular to the laser beam, where it is focused into a grating spectrograph (MS257, Oriel) equipped with a single-photon counting photomultiplier (Burle,

Sample	$\langle d \rangle$ ($e\,a_0$)	λ (nm)	τ (ns)
free Cs atom, exp. [16]	-4.4978(61)	894.6	34.82
free Cs atom, calc.	-4.41	894.3	36.2
Cs in bcc He, exp.	-4.61	884.2	32.5
Cs in bcc He, calc., without cavity effect	-4.33	882.5	35.8
Cs in bcc He, calc., with cavity effect	-4.56	885.9	32.8

TABLE 5.1: Calculated and experimental reduced dipole matrix elements $\langle d \rangle = \langle 6P_{1/2} \| d \| 6S_{1/2} \rangle$ in atomic units, transition wavelengths and lifetimes of the free Cs atom and of Cs in a spherical bubble. All results, except for the one in the first line are from the present work.

model C31034, cooled to -30° C, 2.5 ns rise-time) mounted after the output slit. The spectrograph suppresses very efficiently scattered laser light, whose wavelength differs by 30 nm from the fluorescence wavelength. The excitation laser is triggered by a pulse generator, a delayed pulse of which is used as a start signal for a time-to-amplitude converter (TAC). The photomultiplier (PM) is operated in the single-photon counting mode. The PM pulses are amplified and analyzed by a constant fraction discriminator (CFD), whose logic output serves as the stop signal for the TAC. The TAC delivers analogue voltage pulses whose amplitudes are proportional to the time delay between the start and stop pulses. A histogram of the amplitudes of the TAC pulses is recorded in 0.4 ns wide time bins using a multiscaling card in a personal computer. This histogram directly reflects the time dependence of the fluorescence intensity. In order to avoid pile-up effects, the light intensity was attenuated in an appropriate way so that on average less than one photon was detected per excitation pulse.

All measurements were done at 1.5 K or 1.6 K, in the pressure range of 26 - 38 bar, both in the bcc and hcp phase of solid He.

5.3.2 Life time measurements

A typical time trace of the fluorescence is shown in Fig. 5.3 (a), while Fig. 5.3 (b) shows the shape of the laser pulse recorded with the same system. The rise of the fluorescence during the laser pulse is taken into account in the subsequent analysis by fitting a convolution of an exponential decay with the laser pulse shape to the data. In another analysis we determined the decay time by fitting an exponential decay curve to the tail of the signal, where the excitation of Cs atoms no longer interferes with their

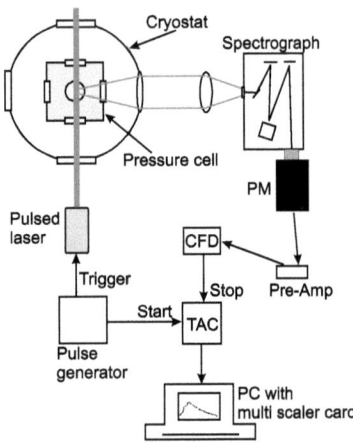

FIGURE 5.2: Setup for the lifetime measurements. The repetition rate of the excitation laser is 10 kHz. Details are given in the text.

decay. The two methods yield lifetimes that coincide within 2 %, which is on the order of the statistical errors. The values shown in Fig.(5.4) are from the analysis with the convolution method.

We studied the pressure dependence of the lifetime τ of the Cs $6P_{1/2}$ state in the range including both bcc (at 26 bar) and hcp (up to 38 bar) phases of solid He. The results are shown in Figure 5.4, which also shows lifetimes of the Cs $6P_{1/2}$ measured in superfluid He taken from [9]. The authors of that reference have measured lifetimes of the lowest $nP_{1/2}$ states of Rb and Cs in condensed He from the saturated vapor pressure up to the solidification

5.4 Discussion

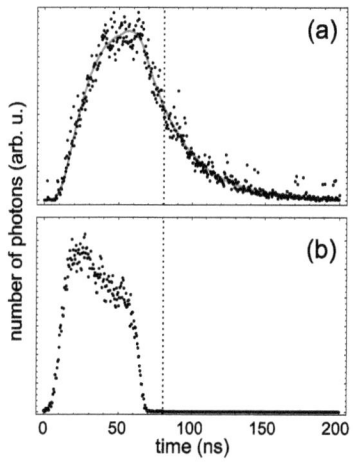

FIGURE 5.3: Histogram of arriving photon delay times (dots). (a) shows the decay curve of the $6P_{1/2}$ state of cesium. The solid (red) line is a fit to the experimental points. The fit function is a convoluting of the excitation pulse shape, shown in (b), with an exponential decay $\exp(-t/\tau)$. The vertical dotted line shows the point, where the excitation pulse is off, and from whereon the decay curve in (a) is a pure exponential.

point at 25 bar. For Cs they found a lifetime of approximately 32.3 ns, which is nearly independent on the helium pressure, while the lifetimes in Rb decrease in a dramatic way with increasing pressure, leading to a complete quenching of the fluorescence at the solidification pressure. Here we find that the lifetimes of Cs in the bcc phase coincide with the (pressure independent) values in liquid He. When increasing the He pressure beyond the bcc phase we first observe a pronounced jump by 3.2(9) ns to shorter lifetimes at the bcc–hcp phase transition, and then a further decrease of the lifetime with increasing He pressure at a rate of -0.30(6) ns/bar.

5.4 Discussion

5.4.1 Lifetimes in the bcc phase

The experimental lifetime values obtained in the bcc phase of the He matrix are close to the ones measured previously in liquid He. Our theoretical calculation in the frame of the SBM (presented in

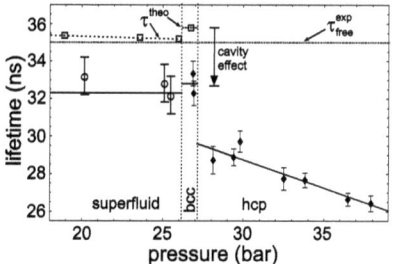

FIGURE 5.4: Pressure dependence of the Cs $6P_{1/2}$ lifetime in solid and liquid He. The open circles are experimental values from [9] and the horizontal solid line through these points represents their average value including points down to zero pressure not shown in the graph. The vertical dotted lines mark different phase boundaries of condensed He. The horizontal dotted line at 34.8 ns indicates the lifetime of the free Cs atom. The open squares are theoretical lifetimes obtained in this work by the SBM. The cavity effect is discussed in the text.

section 5.2) yields a value τ^{theo} in the bcc phase of 35.8 ns at $T = 1.5$ K and $p = 26.6$ bar. If we apply the correction due to the interaction of the atomic dipole with the surrounding dielectric He bulk (cavity effect, discussed in the theory section we obtain $\tau = 35.8/1.086 = 33.0$ ns, in excellent agreement with the experimental value $\tau = 32.8(9)$ ns (average of the two data points in bcc). Because of the pressure independence of the lifetimes, this correction (the index of refraction of liquid He ($n_{\text{He}} = 1.0365$) is close to the one of solid bcc He) also leads to an agreement between the experimental lifetimes [9] and our theoretical values in superfluid He. However, this good agreement has to be taken with some caution, considering that our wavefunctions for the free Cs atom reproduce the experimental lifetime of the free atom $\tau^{\text{exp}}_{\text{free}} = 34.8$ ns only at a level of 4%. Nevertheless it seems that we have strong evidence that the alteration by the dielectric cavity gives an important contribution to the lifetime of excited Cs atoms in superfluid and bcc solid He.

5.4.2 Lifetimes in the hcp phase

As can be seen in Fig. 5.4 the pressure dependence of the lifetime makes an abrupt jump of -3.2(9) ns at the bcc–hcp phase transition. The SBM cannot account for this discontinuity. Moreover, the lifetimes show a pronounced pressure dependence in the hcp phase, another fact which the SBM cannot explain; the SBM calculations predict a change of τ by approximately -0.5 ns when increasing the pressure from 27 to 34 bar, whereas the experimental lifetime changes by approximately -1.6 ns over that range.

We assign the increased deexcitation rate in the hcp phase to the morphological change of the atomic bubble structure from a spherical shape in the bcc matrix to an elliptically deformed bubble in the uniaxial hcp matrix. Experimental evidence for such bubble deformations were presented earlier based on optical and magnetic resonance experiments (a review of those experiments is given in [7]). It seems that this structural change opens a new deexcitation channel for the $6P_{1/2}$ state which most likely consists in the formation of Cs*He$_2$ and Cs*He$_N$ ($N = 6$ or 7) exciplexes. Recently we have performed a detailed spectroscopic study of such exciplexes [4] which fluoresce at a strongly red-shifted wavelength, not detected in our present set-up.

Cs-He exciplex formation has not been observed in liquid nor in bcc He when the Cs atoms are excited at the D_1 transition, while Rb-He exciplexes were observed under similar conditions in Rb-doped liquid He [20]. For both Rb and Cs the exciplex formation after D_1 excitation involves the tunnelling of He atoms through a potential barrier V_0 [4]. It is known [9, 4] that the potential barrier is lower for Rb than for Cs, so that the density of pressurized liquid He is sufficient for a significant formation of exciplexes during the lifetime of the $5P_{1/2}$ state. This is nicely illustrated by the He pressure-dependent fluorescence quenching of Rb in HeII, resulting in a complete disappearance of the Rb fluorescence even before the liquid-bcc phase transition is reached [9]. An increase of the pressure reduces the bubble size and thereby lowers the effective barrier height, so that the tunnelling probability increases with He pressure. This general feature explains the pressure dependent decrease of atomic Rb fluorescence in liquid He and of atomic Cs fluorescence in hcp solid He. With the lower barrier height of Rb the exciplex formation thus starts already in the superfluid phase, while with Cs one needs the higher density, and more importantly, the change of symmetry of the trapping site in the solid hcp phase for observing the quenching by exciplex formation.

5.4.3 Analysis

The diagram in Fig. 5.5 shows the two concurring deexcitation channels of the $6P_{1/2}$ state after D_1 excitation at 850 nm. The two processes can be modeled in terms of decay rates by writing the total

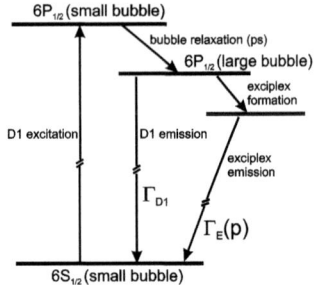

FIGURE 5.5: Decay channels of the $6P_{1/2}$ state of Cs in solid He following atomic D_1 absorption at 850 nm. The excitation takes place in a small bubble which relaxes towards a larger bubble with a lower energy. The excited atom in the larger bubble can decay to the ground state either via atomic D_1 emission (at 880 nm) or via exciplex formation and subsequent exciplex fluorescence. Figure is not to scale with the energy levels.

$6P_{1/2}$ state decay rate in hcp as

$$\Gamma_{6P_{1/2}} = \Gamma_{D1} + \Gamma_E(p) \quad (5.3)$$
$$= \Gamma_{D1} + kW_T, \quad (5.4)$$

where $\Gamma_{D1} = \tau_{D1}^{-1}$, $\Gamma_E(p)$ is the exciplex formation rate, and k is the rate of collisions with the potential barrier for exciplex formation [4]. The probability (Gamov factor) to tunnel through the barrier is

$$W_T = \exp\left[-2b\sqrt{\frac{2m}{\hbar^2}(V_0 - E)}\right] \quad (5.5)$$

where we have assumed for simplicity a square potential of height V_0 and width b. If we assume that the effective barrier height $V_0 - E$ and the barrier width b have a small linear variation with helium pressure p_{He}, the decay rate can be written as

$$\Gamma_{6P_{1/2}} = \Gamma_{D1} + \Delta\Gamma_{bcc-hcp} \exp\left[\alpha(p_{He} - p_{bcc-hcp})\right]$$
$$\approx \Gamma_{D1} + \Delta\Gamma_{bcc-hcp} + \beta(p_{He} - p_{bcc-hcp}), \quad (5.6)$$

where $p_{bcc-hcp}$ is the pressure at the bcc-hcp phase transition, and where $\Delta\Gamma_{bcc-hcp}$ is the jump of the decay rate at the bcc-hcp phase boundary, if we take the radiative lifetimes in the bcc and in the hcp phases to be identical. The experimental value of $\Delta\Gamma_{bcc-hcp}$ is $+3.2(9)\cdot 10^6$ s^{-1}. Figure 5.6 shows the total decay rate of the $6P_{1/2}$ state as a function

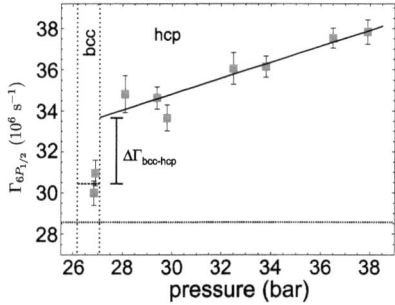

FIGURE 5.6: Decay rate of the $6P_{1/2}$ state as a function of pressure. The solid line is a fit to data points using the model of Eq. 5.6. The horizontal dotted line indicates the radiative decay rate of the free Cs atom.

of pressure. The function of Eq. 5.6 is fitted to the data points and shown as a solid line. The fit yields $\beta = 0.38(7) \cdot 10^6 \mathrm{s}^{-1} \mathrm{bar}^{-1}$.

From Eq. 5.6 we can derive the branching ratio to the exciplex channel, i.e., the probability for the $6P_{1/2}$ state to form an exciplex to be

$$W_{\text{exci}} = \frac{\Gamma_E(p)}{\Gamma_E(p) + 1/\tau_{D1}} \quad (5.7)$$
$$= \frac{\Delta\Gamma_{\text{bcc-hcp}} + \beta(p_{\text{He}} - p_{\text{bcc-hcp}})}{\Delta\Gamma_{\text{bcc-hcp}} + \beta(p_{\text{He}} - p_{\text{bcc-hcp}}) + 1/\tau_{D1}}.$$

Figure 5.7 shows the pressure dependence of the branching ratio W_{exci}.

As discussed in Sect. 5.4.2 the barrier height V_0 prevents exciplex formation in the bcc phase. The jump of the exciplex formation rate at the bcc-hcp phase transition can thus be explained by a lowering of the effective potential barrier. This effect can be explained by the slightly smaller bubble size and morphological change of the bubble shape in hcp by which the He atoms at the bubble interface are brought closer to the top of the barrier.

The absolute value of the effective barrier height $V_0 - E$ cannot be inferred from the present data since we do not know the collision rate with the barrier, which depends on the oscillation frequency of the interface atoms, nor details of the barrier shape. As discussed in [4] the ro-vibrational structure of the final bound exciplex allows, in principle, for a resonant tunnelling to occur, whenever E matches a ro-vibrational eigenstate of the exciplex. Such processes might show up as resonant changes in the pressure dependence of the atomic lifetimes.

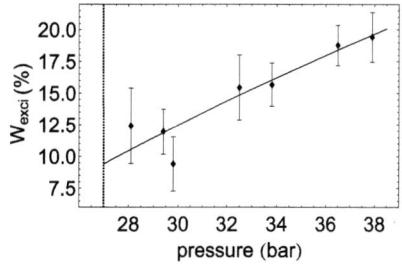

FIGURE 5.7: Exciplex formation probability W_{exci} as a function of pressure. The vertical dashed line marks the pressure of the bcc to hcp phase transition.

5.4.4 Summary

We have presented a detailed experimental study of the pressure dependence of the Cs $6P_{1/2}$ state lifetime in bcc and hcp solid He. The standard bubble model was used to estimate the lifetime of the $6P_{1/2}$ state in superfluid He and in bcc solid He. It reproduces well the pressure independence of the lifetimes in pressurized superfluid He and in the bcc phase. The absolute value predicted by the model is 9% larger than the measured one. We explain this discrepancy by a reduction of the lifetime due to the interaction of the atomic dipole with its mirror images in the dielectric (cavity effect). In the uniaxial hcp phase the observed pressure dependence of the lifetime can be explained by the opening of an additional deexcitation channel which consists in the formation of Cs*He$_2$ and Cs*He$_N$ ($N = 6$ or 7) exciplexes. We suggest that the jump of the lifetime at the bcc-hcp phase transition may be explained by the structural change of the atomic bubble in hcp ^{4}He. The experimental data were used to infer the pressure dependent probability of exciplex formation.

Acknowledgments

This work was supported by the grant number 200021-111941/1 of the Schweizerischer Nationalfonds.

References

[1] T. Kinoshita, K. Fukuda, Y. Takahashi, and T. Yabuzaki. *Phys. Rev. A*, 52(4):2707, 1995.

[2] T. Kinoshita, K. Fukuda, and T. Yabuzaki. *Phys. Rev. B*, 54(9):6600, 1996.

[3] S. Kanorsky, A. Weis, M. Arndt, R. Dziewior, and T.W. Hänsch. *Z. Phys. B*, 98:371, 1995.

[4] P. Moroshkin, A. Hofer, D. Nettels, S. Ulzega, and A. Weis. *J. Chem. Phys.*, 124:024511, 2006.

[5] F. Stienkemeier, J. Higgins, C. Callegari, S. I. Kanorsky, W. E. Ernst, and G. Scoles. *Z. Phys. D*, 38:253, 1996.

[6] J. Peter Toennies and Andrei F. Vilesov. *Annu. Rev. Phys. Chem.*, 49:1–41, 1998.

[7] P. Moroshkin, A. Hofer, S. Ulzega, and A. Weis. *Fiz. Nizk. Temp.*, 32:1297–1319, 2006. (Low Temp. Phys. 32(11), 981-998 (2006)).

[8] Q. Hui, J. L. Persson, J. H. M. Beijersbergen, and M. Takami. *Z. Phys. B*, 98:353, 1995.

[9] T. Kinoshita, K. Fukuda, T. Matsuura, and T. Yabuzaki. *Phys. Rev. A*, 53(6):4054, 1996.

[10] S. I. Kanorsky, M. Arndt, R. Dziewior, A. Weis, and T. W. Hänsch. *Phys. Rev. B*, 50(9):6296, 1994.

[11] J. Pascale. *Phys. Rev. A*, 28(2):632, 1983.

[12] P. Gombas. Springer-Verlag Berlin-Gottingen-Heidelberg, 1956.

[13] A. Weis. *Atoms in nanocavities*, volume 314. Kluwer Academic Publisher, 1995.

[14] A. Hofer, P. Moroshkin, S. Ulzega, and A. Weis. 2007. to be published.

[15] D. W. Norcross. *Phys. Rev. A*, 7:606, 1973.

[16] R. J. Rafac, C. E. Tanner, A. E. Livingston, K. W. Kukla, H. G. Berry, and C. A. Kurtz. *Phys. Rev. A*, 50(3):1976, 1994.

[17] R. J. Rafac, C. E. Tanner, A .E. Livingston, and H. G. Berry. *Phys. Rev. A*, 60(5):3648, 1999.

[18] V. V. Klimov and V. S. Letokhov. *Chem. Phys. Lett.*, 301:441, 1999.

[19] M. Arndt, R. Dziewior, S. Kanorsky, A. Weis, and T. W. Hänsch. *Z. Phys. B*, 98:377, 1995.

[20] K. Hirano, K. Enomoto, M. Kumakura, Y. Takahashi, and T. Yabuzaki. *Phys. Rev. A*, 68:012722, 2003.

Chapter 6

Paper IV:
Rb*He$_n$ exciplexes in solid ^{4}He

This paper analyzes the different emission lines of excited Rb atoms implanted in solid ^{4}He. We have studied the emission following D_1 and D_2 excitation. In both cases only a very weak atomic emission is observed. The main decay channel is the Rb*He$_6$ exciplex formation. The theoretical exciplex model is used to identify the observed emission lines.

My main contributions to the work were:

- Developing software (C++ builder) to control the spectrograph and to read out signals from a photodiode. The signals from the photodiode are recorded with an ADC card in a PC. A new InGaAs photodiode was installed to extend the spectral range of the detection system up to 1700 nm. An appropriate mount for the InGaAs photodiode including lenses was built to collect a maximum of fluorescence light. The Ti:Sa laser was aligned and calibrated with a new set of mirrors in the laser cavity. This laser was used to excite the Rb atoms and the new mirror set allowed us to cover the spectral range of the Rb absorption bands in solid He (blue shifted with respect to the free atom).

- Recording the data together with my colleagues P. Moroshkin and S. Ulzega.

- Analyzing the experimental data with Mathematica.

- Applying the model developed for the Cs-exciplex system to the Rb-exciplex system. Comparing the experimental data to the exciplex model developed by D. Nettels (main), P. Moroshkin and myself, thereby identifying the observed emission lines.

- Producing figures, graphs and text for the paper.

Rb*He$_n$ exciplexes in solid ^{4}He

A. Hofer[1], P. Moroshkin[1], D. Nettels,[1], S. Ulzega[1] and A. Weis[1]

[1]*Département de Physique, Université de Fribourg, Chemin du Musée 3, 1700 Fribourg, Switzerland*

Published in Phys. Rev. A **74**, 032509 (2006).

Abstract: We report the observation of emission spectra from Rb*He$_n$ exciplexes in solid ^{4}He. Two different excitation channels were experimentally identified, viz., exciplex formation via laser excitation to the atomic $5P_{3/2}$ and to the $5P_{1/2}$ levels. While the former channel was observed before in liquid helium, on helium nanodroplets and in helium gas by different groups, the latter creation mechanism occurs only in solid helium or in gaseous helium above 10 K. The experimental results are compared to theoretical predictions based on the extension of a model, used earlier by us for the description of Cs*He$_n$ exciplexes. We also report the first observation of fluorescence from atomic rubidium in solid helium, and discuss striking differences between the spectroscopic feature of Rb-He and Cs-He systems.

6.1 Introduction

The formation process of alkali-He$_n$ exciplexes, i.e., of bound states of an excited alkali atom with one or more ground state helium atoms, was studied in recent years in superfluid [1, 2] and in solid [3] helium. These studies have given support to earlier proposals [4, 5], which tentatively explained the quenching of atomic fluorescence from light alkali atoms (Li, Na, K) in condensed helium by the formation of alkali-helium exciplexes, whose emission spectra are strongly red-shifted with respect to the atomic resonance lines. Exciplex formation was also studied on the surface of helium nanodroplets [6, 7, 8, 9, 10] and in cold helium gas [11, 1, 2]. Recently we have performed an experimental and theoretical study of the Cs*He$_n$ exciplex formation process in the hcp and bcc phases of solid ^{4}He [12]. A comparison with the results of [1, 2] has revealed that the exciplex formation mechanism in solid helium differs from the one in superfluid helium and in cold helium gas. We concluded that exciplexes in solid helium result from the collective motion of several nearby helium atoms which approach the Cs atom simultaneously, while in liquid and gaseous helium the binding of the helium atoms proceeds in a time sequential way.

The motivation for the present study of the Rb-He system arose from the question whether the collective mechanism is specific for Cs in solid helium, or whether it also holds for other alkali atoms. While the light alkali atoms (Li, Na, K) do not emit resonance fluorescence when excited in condensed helium, atomic cesium fluoresces both in superfluid and in solid helium, when excited on the D$_1$ transition. Rubidium represents an intermediate case, as it was reported [13] to fluoresce in liquid helium when excited on the D$_1$ transition with a yield which is strongly quenched with increasing He pressure. No fluorescence from Rb in solid helium was observed in the past, although it was shown that optically detected magnetic resonance can be used to detect light absorption on its D$_1$ transition [14].

A major difference between cesium and rubidium exciplexes Rb/Cs(A$^2\Pi_{1/2}$)He$_n$ becomes apparent from Fig. 6.1 which shows the calculated binding energies ϵ_b(Rb) (ϵ_b(Cs)) of the exciplexes as a function of the number n of bound helium atoms for Rb (Cs). For Cs only exciplexes with 5, 6 and 7 helium atoms have their energy below the dissociation limit and are therefore stable, while for Rb all exciplexes with $n = 1 \ldots 8$ are stable.

For cesium the binding energy has a local mini-

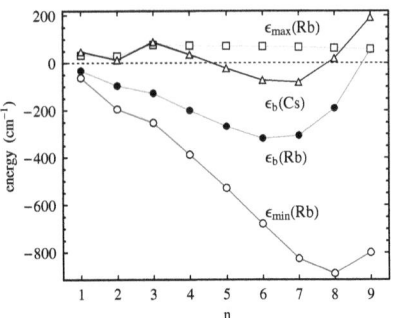

FIGURE 6.1: Calculated energies of Rb(A$^2\Pi_{1/2}$)He$_n$ exciplexes as a function of the number n of attached helium atoms. All energies (defined in Fig. 6.2(b)) are given with respect to the dissociation limit, i.e., the energy of the $5P_{1/2}$ state of free Rb. Shown here are the depths of the potential wells ϵ_{min}(Rb) (open circles), the barrier heights ϵ_{max}(Rb) (open squares) and the binding energies ϵ_b(Rb) (solid dots). The binding energies ϵ_b(Cs) (open triangles) of Cs exciplexes from [12] are shown for comparison.

mum for $n = 2$ (quasi-bound complex) and there is a potential barrier that hinders the formation of exciplexes with more than 2 helium atoms in a sequential manner. As evidenced by the measurements of [1] the Cs*He$_{n=2}$ exciplex is therefore the largest complex that can be formed by a sequential attachment of He atoms. Larger complexes can only be formed in a collective way, which becomes possible in pressurized solid helium [3]. The largest stable complex will be the one with the lowest binding energy. For Rb all the exciplexes with $n = 1 \ldots 8$ are stable, so once the Rb*He$_{n=1}$ exciplex is created all larger complexes can be formed with high probability by the sequential filling of the helium ring until the state with the lowest binding energy is reached. In helium environments with lower densities than pressurized solid helium the time intervals between successive attachments is long enough to permit the exciplex to fluoresce, so that fluorescence from all intermediate exciplexes Rb*He$_{n=1\ldots6}$ can be observed in gaseous helium [2]. The results presented below indicate that in solid He the Rb(A$^2\Pi_{1/2}$)He$_n$ formation process is so rapid that any intermediate configurations have no time to emit fluorescence. For Rb in solid helium one therefore expects that only the most strongly bound Rb*He$_6$ exciplex is formed.

In Sec. 6.2 we review the theoretical model for the description of exciplex spectra developed in [12] and extend it to the Rb-He system. In Sec. 6.3 we introduce the experimental setup and present experimental emission and excitation spectra of rubidium-helium exciplexes. In Sec. 8.6 we compare the experimental results with the theoretical model calculations as well as other experiments and discuss the different decay channels of excited Rb in solid helium.

6.2 Theory

We briefly describe the theoretical approach of our calculation of the Rb*He$_n$ exciplex emission spectra for $n = 1-9$. The model used is an extension of the calculations performed earlier for cesium-helium exciplexes [12, 3] and we shall review only the basic principles and assumptions. We consider only the interaction of the excited Rb atom with the n helium atoms that form the exciplex and neglect the influence of the helium bulk. The largest perturbation comes from the close helium atoms that form the exciplex and it is therefore a good approximation to neglect the helium bulk. The interaction between the Rb atom and one ground state helium atom is described as a sum over semi-empirical pair potentials [15]

$$V_n^{\text{Rb-He}}(r) = \sum_{i=1}^{n} V^{5P}(\mathbf{r_i}), \quad (6.1)$$

where $\mathbf{r_i}$ is the position of the i-th helium atom with respect to the position of the Rb atom. After including the spin-orbit interaction of the Rb valence electron and the helium-helium interaction, $V_n^{\text{He-He}}(r)$, modeled as the sum over interaction potentials [16] between neighboring helium atoms the total interaction Hamiltonian is given by

$$V_{\text{Rb*He}_n}(r) = V_n^{\text{Rb-He}}(r) + V_n^{\text{He-He}}(r) + (2/3)\Delta \mathbf{L} \cdot \mathbf{S}, \quad (6.2)$$

where $\Delta = 237.6 \, \text{cm}^{-1}$ is the fine structure splitting of the rubidium $5P$ state in the free atom. $\mathbf{L}$ is the orbital angular momentum operator and $\mathbf{S}$ the electronic spin operator. Next, the total interaction operator $V_{\text{Rb*He}_n}(r)$ is represented in the basis $|n, L, S\rangle$ and diagonalized algebraically. Exciplexes of two different structures are formed as in the case of cesium-helium exciplexes. When one or two helium atoms are bound the electronic wavefunction has an apple shape with the helium atoms attached in its dimples, whereas for $n > 2$ the electronic wavefunction has a dumbbell shape,

with the bound helium atoms distributed along a ring around the dumbbell's waist. The potential curves leading to the formation of these two classes of structures are represented in Fig. 6.2 using the examples of Rb*He$_2$ and Rb*He$_6$. The potential curves shown represent the r-dependent eigenvalues of the operator $V_{\text{Rb*He}_n}(r)$ of Eq. (6.2). In the same figures we also show the ground state potentials $nV_\sigma^{5S}(r) + V_n^{\text{He-He}}(r)$. We will use the standard spectroscopic notation ($X^2\Sigma_{1/2}$, $A^2\Pi_{1/2}$, $B^2\Pi_{3/2}$ and $C^2\Sigma_{1/2}$) also for complexes with $n > 2$ for simplicity and to be consistent with our previous publications [12, 3] although this notation is strictly speaking only valid for linear molecules.

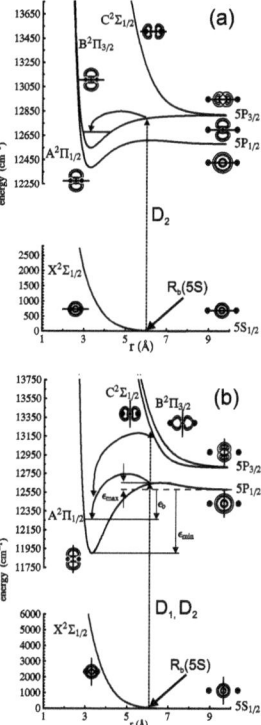

FIGURE 6.2: Adiabatic potentials of the Rb*He$_n$ system: (a) Rb*He$_2$, (b) Rb*He$_6$. The equilibrium bubble radius of the ground state Rb atom is indicated with R$_b$(5S). The energies shown in Fig. 6.1 as a function of the number of bound helium atoms are visualized in (b).

As can be seen from Fig. 6.2(a) the energetically most favorable formation channel for Rb*He$_2$ proceeds via D$_2$ excitation; when two helium atoms approach along the nodal line of the apple-shaped electron distribution of the B$^2\Pi_{3/2}$ state, they are attracted into the potential minimum. When the system is excited on the D$_1$ transition the approaching helium atoms see a repulsive spherical electronic distribution of the Rb atom at large distances with a potential barrier of 29 cm^{-1}. We recall that the corresponding barrier height in cesium is 79 cm^{-1} [12] due to the larger spin-orbit interaction energy in that atom [4]. The approaching helium atoms deform the electronic configuration of the 5P state from spherical to apple shaped.

The exciplexes with $n > 2$ [Fig. 6.2(b)] have no potential well in the B$^2\Pi_{3/2}$ state, which is purely repulsive and which correlates to the 5$P_{3/2}$ atomic state. However, the A$^2\Pi_{1/2}$ state possesses a potential well and a potential barrier. The barrier is associated with the transformation of the electronic wavefunction from spherical to dumbbell-shaped when several helium atoms approach the Rb atom. Exciplexes with $n > 2$ can only be formed in the A$^2\Pi_{1/2}$ state.

The electronic distributions of the rubidium-helium system for the different states at various interatomic separations are illustrated by pictographs in Fig. 6.2. The solid lines represent the quantization axis, which is the internuclear axis for Rb*He$_{n\leq 2}$ and the symmetry axis of the helium ring for the Rb*He$_{n>2}$ complexes, while helium atoms are drawn as filled disks with a radius of 3.5 Å.

In a next step we have calculated the vibrational zero-point energies for all Rb*He$_n$ for $n = 1\ldots 9$. Details of this calculation were discussed in [12] for the case of cesium. Only the lowest vibrational state is considered as higher vibrational states are not populated at the temperature (T=1.5 K) of the experiment. A more detailed discussion about this statement will be given in Sec. 6.4.2. The binding energies ϵ_b(Rb), ϵ_b(Cs), the well depths ϵ_{min}(Rb) and the barrier heights ϵ_{max}(Rb) are shown in Fig. 6.1 for Rb(A$^2\Pi_{1/2}$)He$_{n=1\ldots 9}$.

As a last step we calculate the emission spectra $I(\nu)$ of all Rb*He$_{n=1\ldots 9}$ exciplexes under the Franck-Condon approximation as discussed in [12]. The theoretical emission spectra for Rb(B$^2\Pi_{3/2}$)He$_{n=1,2}$ and for Rb(A$^2\Pi_{1/2}$)He$_{n=6,7}$ are shown in Fig. 6.3.

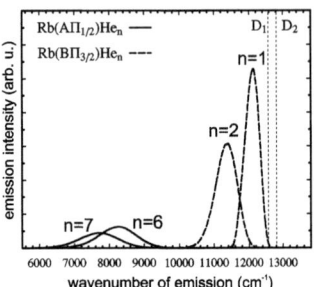

FIGURE 6.3: Calculated emission spectra of Rb(B$^2\Pi_{3/2}$)He$_{n=1,2}$ (dashed lines) and Rb(A$^2\Pi_{1/2}$)He$_{n=6,7}$ (solid lines). The dotted lines indicate the positions of the resonance lines of the free Rb atom.

6.3 Experimental results

6.3.1 Experimental setup

The experimental setup is similar to the one described in our previous publication [3]. A helium crystal is grown at pressures around 30 bar in a pressure cell immersed in superfluid helium at 1.5 K. The matrix is doped with rubidium atoms by laser ablation using a frequency-doubled Nd:YAG laser. The cell has five windows for admitting the ablation beam and the beam of the spectroscopy laser (a tunable cw Ti:Al$_2$O$_3$ laser) and for collecting fluorescence from the sample volume. The fluorescence is dispersed by a grating spectrometer and recorded, depending on the spectral range under investigation, either by a CCD camera (9500...13500 cm^{-1}) or by an InGaAs photodiode (5500...9500 cm^{-1}). We shall refer to these as CCD-spectrometer and InGaAs-spectrometer respectively. With the InGaAs-spectrometer spectra were recorded by a stepwise tuning of the grating, while integral spectra could be recorded with the CCD-spectrometer.

6.3.2 Atomic Bubbles

Defect atoms in solid helium reside in atomic bubbles, whose size and structure can be described by the equilibrium between a repulsive alkali-helium interaction due to the Pauli principle on one hand and surface tension and pressure volume work on the other hand [17, 18, 5]. The interaction with the helium bulk shifts the $5S_{1/2} \to 5P_{1/2}$ (D$_1$) and $5S_{1/2} \to 5P_{3/2}$ (D$_2$) transitions of Rb by approximately 35 nm to the blue with respect to their values (794 nm and 780 nm respectively) in the free atom. This shift of the excitation lines as well as a smaller blue shift of the corresponding emission lines is well described by the bubble model [5, 19]. We have calculated the equilibrium radius of the atomic bubble formed by the $5S_{1/2}$ ground state of the Rb atom to be $R_b = 6$ Å (Fig. 6.2) following the model described in [17, 18]. For the interaction potential between ground state Rb and He atoms we have used the same semi-empirical potentials [15] as for the exciplex model.

It is the close vicinity of the helium atoms in the first solvation shell, together with their large zero point oscillation amplitudes, which form the basis of the efficient exciplex formation in solid helium.

6.3.3 Emission spectra following D_1 excitation

Fig. 6.4 shows the emission spectrum recorded with the CCD-spectrometer following excitation at the D_1 wavelength 13140 cm^{-1} (758 nm). The peak b' at 12780 cm^{-1} represents fluorescence from the atomic $5P_{1/2}$ state. While D_1 atomic fluorescence from Cs in solid helium has been studied and used extensively in the past it was believed that rubidium would not fluoresce on the D_1 transition when embedded in solid helium. This belief was based on the reported quenching of the atomic fluorescence at high pressures in superfluid helium [13]. It should be noted that the Rb-D_1 fluorescence reported here is orders of magnitude weaker than the corresponding line in Cs and could only be detected with long integration times (4 seconds) of the CCD camera, which probably explains why this spectrum was not observed in previous experiments [14].

The apple-shaped exciplexes with one or two bound helium atoms are expected to fluoresce within the spectral range of Fig. 6.4 and the absence of any prominent spectral feature indicates that these complexes are not formed upon D_1 excitation. The sloped background visible in Figs. 6.4 and 6.6 is a strong wing of scattered laser light ($\lambda = 13160$ cm^{-1}). The inset in Fig. 6.4 shows a spectrum which was recorded using a grating with a higher resolution. The excitation laser was shifted by 65 cm^{-1} (still in the D_1 absorption band (Fig. 6.7)) to the blue with respect to the spectrum of Fig. 6.4 to make clear that no D_2 emission can be observed after D_1 excitation. The arrow in the inset indicates the position of the D_2 emission measured after D_2 exciation (peak a in Fig. 6.6).

When exploring the longer wavelength range with the InGaAs-spectrometer we found a very strong fluorescence band (Fig. 6.5) centered at 7420 cm^{-1}, which we assign to Rb*He$_{n>2}$ exciplexes in the $A^2\Pi_{1/2}$ state. This is the first recording of such exciplexes after D_1 excitation in solid He and the proof that the quenching of atomic D_1 fluorescence in rubidium [13] is due to exciplex formation. A similar emission following D_1 excitation has been observed in gaseous He above 10 K [2]. Measurements at lower He gas temperatures and measurements in liquid He at 1.8 K have shown no exciplex formation after D_1 excitation [2]. The question why exciplex formation becomes possible again in solid He will be addressed in Sec. 6.4.5. The dashed and the solid lines in Fig. 6.5 are theoretical emission spectra from Rb*He$_6$ and Rb*He$_7$ respectively. Figure 6.5(b) shows the theoretical curves, shifted such as to make their blue wings coincide with the experimental points. The line shape of the experimental curve is well reproduced by the two theoretical curves. The theoretical curve of the Rb*He$_7$ fits the experimental points better on the low energy side, while on the high energy side both curves are in very good agreement with the experimental spectrum. A small discrepancy is visible on the low energy wing, which can be due to imprecisions of the strongly sloped ground state potential (Fig. 6.2) or to changes of the latter due to the helium bulk. It is a remarkable fact that the fluorescence yield of this exciplex after D_1 excitation in solid helium is larger than after D_2 excitation, while it was not observed at all in superfluid helium. We will come back to this point in Sec. 6.3.

A similar emission at around 7200 cm^{-1} has been seen in liquid helium by the Kyoto group [2] after D_2 excitation and was assigned to the emission by the Rb*He$_6$ exciplex.

6.3.4 Emission spectra following D_2 excitation

Fig. 6.6 shows the emission spectrum, measured with the CCD-spectrometer, when the laser is tuned to the atomic D_2 transition at 13420 cm^{-1} (745 nm).

Four prominent spectral features can be seen in the emission spectrum. The two rightmost peaks (labelled a and b) represent atomic D_2 and D_1 fluorescence respectively. Together with the peak of Fig. 6.4 they constitute the first observation of atomic fluorescence from rubidium in solid helium. The presence of D_1 emission after D_2 excitation is evidence for the existence of a fine structure relaxation channel. We assign the two broader fea-

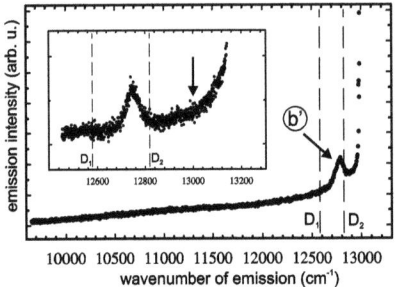

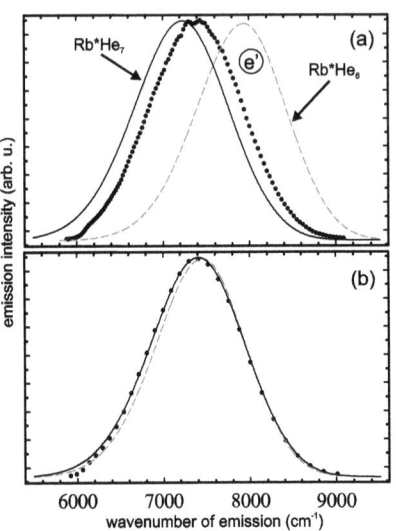

FIGURE 6.4: Measured emission spectrum (dots) recorded with the CCD-spectrometer following D_1 excitation. The dashed vertical lines indicate the D_1 and D_2 lines of the free Rb atom. The peak b' is the fluorescence from the D_1 transition. The inset shows the spectral range around the D-lines recorded with a higher resolution grating and an excitation frequency slightly (65 cm^{-1}) shifted to the blue. The rise on the right side is from scattered laser light. The arrow gives the position at which D_2 emission is detected after D_2 excitation (peak a in Fig. 6.6).

tures c and d peaked at 12400 cm^{-1} and 11800 cm^{-1} respectively to the emission from Rb($B^2\Pi_{3/2}$)He$_1$ and Rb($B^2\Pi_{3/2}$)He$_2$ exciplexes. The solid lines in Fig. 6.6 are the calculated $n = 1$ and $n = 2$ emission spectra of Fig. 6.3 shifted to the blue by Δ_1 and Δ_2 respectively, so that their line centers coincide with the positions of the measured curves. The shifts are probably due to the interaction with the surrounding helium bubble. Note that the two theoretical curves have to be shifted by different amounts in order to match the experimental lines. We have found previously in the Cs-He system [12] that the rate and sign of the pressure shift of exciplex emission lines depend on the number of bound helium atoms.

As with the spectra of Sec. 6.3.3 we have recorded the emission in the region of longer wavelengths with the InGaAs-spectrometer. As a result we find a spectrum, which is identical (same central wavelength and same width) with the one observed with D_1 excitation (Fig. 6.5). This suggests that the emission stems from the same state ($A^2\Pi_{1/2}$) as the emission after D_1 excitation. The population of that state following D_2 excitation is another proof of the existence of a fine structure relaxation mechanism. No other exciplex emission was observed in

FIGURE 6.5: Fluorescence spectrum (dots) following D_1 excitation measured with the InGaAs-spectrometer. The emission band stems from a Rb*He$_{n>2}$ exciplex (e'). An identical emission spectrum was observed after D_2 excitation. (a) The dashed line is a calculated emission spectrum from Rb*He$_6$ and the solid line from Rb*He$_7$. (b) The two theoretical spectra are shifted in order to match the experimental curve.

the spectral range between the Rb*He$_{n>2}$ and the Rb*He$_2$ exciplex emission (peak e' in Fig. 6.5(a) and peak d in Fig. 6.6 respectively).

6.3.5 Atomic and exciplex excitation spectra

The experimental emission spectra presented above were recorded with two fixed excitation wavelengths, chosen such as to maximize the signals of interest. It is of course interesting to investigate how the different spectral features depend on the excitation wavelength. For this we have varied the wavelength of the Ti:Al$_2$O$_3$ laser in discrete steps over the spectral range of 13000...13700 cm^{-1} (~770...730 nm). For every excitation wavelength we have measured the amplitudes of the emission peaks of Figs. 6.4, 6.5 and 6.6.

The top part of Fig. 6.7 shows the excitation spectrum of D_2 fluorescence, which is centered at

6.4 Discussion

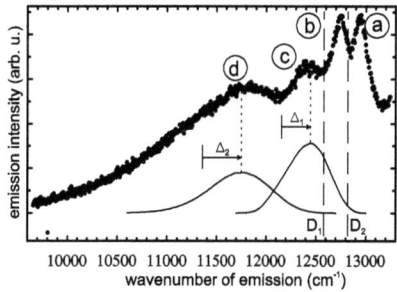

FIGURE 6.6: Fluorescence spectrum (dots) recorded with the CCD-spectrometer following D_2 excitation. The dashed vertical lines indicate the position of the D_1 and D_2 line of the free Rb atom. The following assignments are made to the emission peaks: atomic D_2 fluorescence (a), atomic D_1 fluorescence (b), emission from $Rb(B^2\Pi_{3/2})He_1$ exciplexes (c), and emission from $Rb(B^2\Pi_{3/2})He_2$ exciplexes (d). The solid lines are calculated emission spectra from $Rb(A^2\Pi_{3/2})He_1$ and $Rb(A^2\Pi_{3/2})He_2$ exciplexes. The lines are shifted in order to match the peaks of the experimental curves. $\Delta_1 = 350\,\mathrm{cm}^{-1}$ and $\Delta_2 = 440\,\mathrm{cm}^{-1}$ are the shifts with respect to the calculated positions shown in Fig. 6.3.

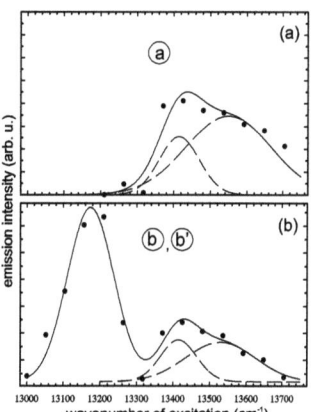

FIGURE 6.7: Excitation spectra of the fluorescence from atomic rubidium: Top: fluorescence analyzing spectrometer set to the D_2 emission line (peak a of Fig. 6.6); bottom: spectrometer set to the D_1 emission line (peaks b' and b of Figs. 6.4 and 6.6). The dashed lines are Gaussians whose sum (solid line) was fitted to the data.

$13460\,\mathrm{cm}^{-1}$ (743 nm). One sees clearly that this fluorescence can only be produced by D_2 excitation. The lower part of Fig. 6.7 shows the excitation spectrum of D_1 fluorescence. It consists of two absorption bands centered at $13180\,\mathrm{cm}^{-1}$ and $13460\,\mathrm{cm}^{-1}$ respectively, which corresponds to excited states correlating with the atomic $5P_{1/2}$ and $5P_{3/2}$ levels respectively. D_1 fluorescence can thus be produced directly via D_1 excitation or via D_2 excitation combined with a J-mixing interaction due to the alkali-helium interaction. The D_1 absorption band is slightly asymmetric with a longer wing on the low energy side. This feature has been observed before in Cs [5]. The D_2 absorption band measured for both D_1 and D_2 fluorescence, has a double peaked-structure. The scarce number of data points is well fitted by a superposition of two Gaussians separated by about $125\,\mathrm{cm}^{-1}$. This splitting of the D_2 excitation lines of cesium and rubidium in superfluid helium has been explained before in terms of a dynamic Jahn-Teller effect due to quadrupolar bubble-shape oscillations which lift the degeneracy of the $P_{3/2}$ state [20].

Fig. 6.8 shows the excitation spectra of the exciplex lines c, d, and e' of Figs. 6.5 and 6.6. As the $Rb^*He_{1,2}$ exciplexes can only be observed after D_2 excitation (Fig. 6.8 c, d) we conclude that these apple-shaped complexes are formed in the $B^2\Pi_{3/2}$ state. The D_1, D_2 and $Rb(B^2\Pi_{3/2})He_{1,2}$ emission lines are very weak and of similar amplitude. The bottom spectrum (e') represents by far the strongest signal that comes from the $Rb(A^2\Pi_{1/2})He_{n>2}$ exciplex which can be excited by either D_1 or D_2 radiation. Its emission line is about 100 times stronger than the other lines. This result is in strong contrast with the emission of the corresponding cesium exciplex, $Cs(A^2\Pi_{1/2})He_{n>2}$, in solid helium, for which the emission after D_1 excitation is very weak [12]. The double-peaked structure of the D_2 excitation spectrum is not well resolved for the $Rb^*He_{1,2}$ exciplexes. It was observed before for Cs^*He and Rb^*He exciplexes on superfluid helium nanodroplets [6, 8].

6.4 Discussion

6.4.1 Atomic lines

The assignment of the atomic D_1 and D_2 excitation and emission lines is unambiguous. The excitation lines are blue-shifted by approximately $600\,\mathrm{cm}^{-1}$,

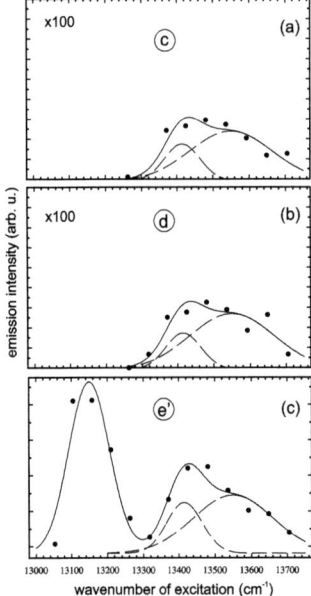

FIGURE 6.8: Excitation spectra of the fluorescence from Rb*He$_n$ exciplexes (dots) with the fluorescence spectrometer tuned to emission from Rb*He$_1$ c, Rb*He$_2$ d, and Rb*He$_{n_{max}}$ e'. The solid lines are Gaussian fits. The signals in the spectrum e' is approximately two orders of magnitude larger than the ones of c and d and than the atomic signals from Fig. 6.7.

while the emission lines are shifted by only 65 cm^{-1} with respect to the free atomic transitions. These shifts (except that of the D$_2$ emission) have been studied in superfluid helium [18] and are well described by the bubble model. The blue shift results from the interaction with the bulk helium, which is less pronounced in the emission process as the latter occurs in a bubble of larger size [5].

As already mentioned, excitation at the D$_1$ transition leads to emission on the D$_1$ line only, while excitation at the D$_2$ line leads to emission on both the D$_1$ and the D$_2$ lines. It should be noted here that in liquid He [18] even under D$_2$ excitation one can only observe D$_1$ emission. We also recall that in Cs-doped condensed He D$_2$ emission is absent in both in the liquid [18] and in the solid [12] phase. The absence of the D$_2$ emission from heavy alkali

atoms in condensed He is explained [1, 2, 12] by the very efficient formation of alkali-helium exciplexes - a general phenomenon observed in the present study as well. We will return to this point in Sec. 6.4.3.

6.4.2 Apple-shaped Rb($B^2\Pi_{3/2}$)He$_{1,2}$ exciplexes

As one can see in Fig. 6.2(a), one or two helium atoms approaching the apple-shaped atomic 5P$_{3/2}$, m$_J$ = ±3/2 state do not experience a potential barrier on their way to the potential well of the $B^2\Pi_{3/2}$ state. The formation process of Rb*He$_1$ and Rb*He$_2$ exciplexes is therefore straightforward after D$_2$ excitation. Note that the potential diagram for Rb*He$_1$ is similar to the one for Rb*He$_2$, shown in Fig. 6.2, with the difference that it has a reduced potential well depth. The Rb*He$_{1,2}$ exciplex emission line following D$_1$ excitation is not observed because only the largest exciplex is formed as discussed in Sec. 6.4.5.

Emission spectra very similar to the one in Fig. 6.6 have been previously observed in gaseous He below 2.1 K [2] and in Rb-doped He droplets [8]. The authors of [2] and [8] assigned their observations to the emission of several vibrational states of the Rb($B^2\Pi_{3/2}$)He$_1$ and Rb($A^2\Pi_{1/2}$)He$_1$ exciplexes. Their calculations of emission spectra support this assignment. However, we believe that in solid He at 1.5 K only the lowest vibrational state is populated and that we observe indeed the emission from two different exciplexes. The reasons are the following. The authors of [2, 1] have shown that the higher vibrational states of the Rb*He$_1$ and Cs*He$_1$ exciplexes are only populated at low He gas densities. For higher densities, especially in liquid He the collision-induced relaxation rate increases and only the lowest vibrational state is populated. The same mechanism should be even more efficient in solid He.

Our assignment of the peak c at 12400 cm^{-1} in Fig. 6.6 to the emission from the lowest vibrational state of Rb($B^2\Pi_{3/2}$)He$_1$ agrees well with the experimental and theoretical results of [2, 8], which place it at 12000 cm^{-1}. Our measurements of the Cs($B^2\Pi_{3/2}$)He$_2$ exciplex [12] demonstrated a pressure-dependent blue shift with a rate of 10 cm^{-1}/bar and a sudden jump of 100 cm^{-1} at the bcc-hcp phase transition. Assuming similar shifts for the Rb*He$_1$ exciplex, we can estimate the difference between the spectral position in liquid He at SVP and in our experiment at 30 bar to be on the order of 400 cm^{-1}.

The position of the peak d at $11800\,\text{cm}^{-1}$ is in good agreement with the position of the lowest vibrational state of $\text{Rb}(A^2\Pi_{1/2})\text{He}_1$ observed (and predicted) in [2, 8] at $11800-11900\,\text{cm}^{-1}$. However, as we discuss in Sec. 6.4.5 only the $\text{Rb}(A^2\Pi_{1/2})\text{He}_n$ exciplex with $n = n_{max}$ emits fluorescence in solid He and we expect the spectral position to be shifted with respect to the measurements in He gas. Therefore we assign this peak to the emission of $\text{Rb}(B^2\Pi_{3/2})\text{He}_2$, which according to our model is stable and should be present in the emission spectrum. Applying the same estimation of the pressure shift as described above one can expect that the emission of this complex in liquid He would be at $11400\,\text{cm}^{-1}$, whereas the calculation of [2] places it at $10900\,\text{cm}^{-1}$. At present we can not explain this discrepancy.

Why the $\text{Rb}(B^2\Pi_{3/2})\text{He}_1$ exciplex is formed in solid He while the correspondent exciplex is not observed in Cs-doped solid He [12] is a more difficult question. We suggest a speculative interpretation in the following section.

6.4.3 Diatomic bubble

We next address the striking difference in the structure of the emission of Rb-doped solid He (present study) compared to Cs-doped condensed (liquid or solid) He [1, 12, 18]. More precisely, in the present study, under D_2 excitation we obtain D_1 and D_2 atomic Rb emission lines plus the emission of two apple-shaped exciplexes, whereas only D_1 atomic emission and one apple-shaped exciplex were observed in similar experiments with Cs.

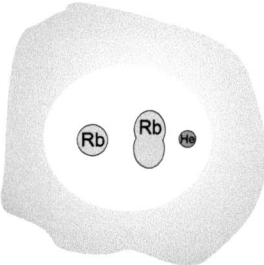

FIGURE 6.9: Sketch of a diatomic bubble after photodissociation of a Rb_2 dimer. A spherical ground state Rb atom and an apple-shaped excited Rb atom share one bubble. Helium atoms can approach only from one side, thus formation of the Rb*He$_1$ exciplex and its fluorescence becomes possible.

In solid He the absorption and emission lines of atoms or molecules are shifted with respect to the value of the free species due to the interaction with the surrounding He bulk. The shift of the atomic lines is well understood in the framework of the bubble model. For Rb in solid He we have the particular situation that the atomic D_2 absorption line overlaps with an dissociative band ($(1)^3\Sigma_u \to (1)^3\Pi_g$) of the Rb_2 dimer [21] also present in the same sample. We speculate that after dissociation two Rb atoms (one in the ground and one in the excited state) share one bubble as shown by a sketch in Fig. 6.9. The excited Rb atom has an apple-shaped orbital and can bind two He atoms but the ground state atom inhibits one binding site. This situation leads to the formation of the $\text{Rb}(B^2\Pi_{3/2})\text{He}_1$ exciplex. At the same time the larger bubble perturbs the excited Rb atom less than the more compact single-atomic bubble, which reduces the quenching efficiency and thus results in measurable D_2 emission.

It is natural to expect Cs_2 dimers to be present in our experiments with Cs-doped solid He. In a separate study [22] we have confirmed this expectation, however, we have found that the photodissociation spectrum of Cs_2 has no overlap with the absorption lines of atomic Cs. The fluorescence spectrum recorded upon photodissociation of Cs_2 molecules at 670 nm ($14925\,\text{cm}^{-1}$) is presented in Fig. 6.10, where one can see both D_1 and D_2 atomic lines together with the $\text{Cs}(B^2\Pi_{3/2})\text{He}_2$ exciplex and the much weaker, but still distinguishable $\text{Cs}(B^2\Pi_{3/2})\text{He}_1$ exciplex. The solid lines represent the exciplex emission spectra calculated using the same approach as described in [12] and in the present paper.

6.4.4 Dumbbell-shaped $\text{Rb}(A^2\Pi_{1/2})\text{He}_{n>2}$ exciplexes

The emission line shown in Fig. 6.5 has the longest wavelength of all observed spectral lines and originates thus from the lowest-lying bound state, i.e., the $A^2\Pi_{1/2}$ state of Fig. 6.2(b). Note that all Rb*He$_{n>2}$ exciplexes have similar potential curves with potential wells increasing with n. All of these structures have the shape of dumbbells, with the helium atoms bound around their waists [12]. Fig. 6.5 also shows the calculated line shapes of the emission from Rb*He$_6$ and Rb*He$_7$. Disregarding shifts of the line centers the theoretical line shapes match the experimental spectrum quite well. The good matching of the line width in particular indicates that this emission is from a single exciplex species

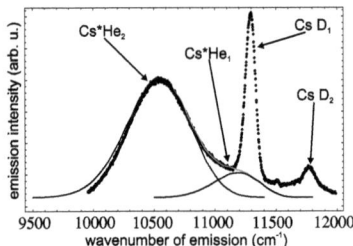

FIGURE 6.10: Spectrum (dots) observed upon photodissociation of the Cs_2 dimer. One can clearly identify the Cs D_1, D_2 and Cs^*He_2 emission lines. The Cs^*He_1 emission is very weak and results in a broadening of the Cs^*He_2 emission line. The black solid lines are calculated emission lines for the two smallest Cs exciplexes. They are shifted in order to fit the data points. The red line is a superposition of the two calculated lines.

with a specific number of bound helium atoms and that it does not come from a superposition of different exciplexes. The shift of the lines is most likely due to the interaction with the helium bulk, which was not taken into account in our calculation. It is difficult to estimate whether the bulk shifts the line to the blue or to the red. One can therefore not assign the observed emission band to Rb^*He_6 or Rb^*He_7 in an unambiguous way. The calculated binding energies $\epsilon_b(Rb)$ (Fig. 6.1) show that the complex with 6 helium atoms has the lowest binding energy and is therefore the most stable exciplex. Observations in liquid He [2] confirm this prediction. However, the exact calculation of the energy of the lowest lying bound state involves a precise quantitative treatment of its oscillatory degrees of freedom. In [12] we have described in detail how we calculate these oscillation energies. There is an uncertainty in the calculated binding energies due to the simplified assumptions we made. An additional uncertainty comes from the semi-empirical pair potentials [15]. For big exciplexes like the Rb^*He_6 every uncertainty in the potential will be amplified because of the additive contribution of the n helium atoms discussed in Sec. 6.2. This can change the position and the depth of the well in the excited state. To all of this adds the effect of the helium bulk, which was not treated so far. The following arguments support the Rb^*He_6 to be the structure observed. It has the minimal binding energy and the corresponding Cs exciplex line is shifted to lower wave numbers with increasing pressure [12].

Assuming the same tendency for the Rb exciplex brings the spectral position of Rb^*He_6 into better agreement with the experimental curve (Fig. 6.5). On the other hand the line shape of the calculated Rb^*He_7 fits better to the data. Therefore we can not conclude which exciplex is the one observed in the experiment.

6.4.5 Formation of dumbbell-shaped $Rb(A^2\Pi_{1/2})He_{n>2}$ exciplexes

The radius of the bubble formed by the rubidium ground state has an equilibrium radius R_b of 6 Å, which is smaller than the corresponding radius for cesium. The excitation process is a Franck-Condon transition to the $5P$ state during which the radius does not change.

The D_1 excitation starting at $R_b(5S)=6$ Å ends at the left of the potential barrier of the $A^2\Pi_{1/2}$ state so that the exciplex is easily formed by helium atoms dropping into the well. Note that for cesium in solid helium the corresponding transition ends on the right side of the potential barrier in the excited state [12]. In that case the helium atoms have to tunnel through the potential barrier in order to form the exciplex. This explains why exciplex emission of Cs in solid helium after D_1 excitation is much weaker than after D_2 excitation, while for Rb the opposite holds. It also explains why no emission from Rb exciplexes after D_1 excitation could be observed in gaseous (below 10 K) and in liquid helium environments [2] in which the helium atoms are, on average, further away from the Rb atom and where the excitation thus ends at the right of the potential barrier. Under those conditions the exciplex formation is strongly suppressed as the helium atoms have to tunnel through the potential barrier to form the exciplex. This tunneling occurs at a rate which is smaller than the exciplex lifetime. The same is true for Rb on He droplets, where no exciplex was observed after D_1 excitation [10]. The authors of [10] estimated the tunneling time to be about 500 ns, much longer than the lifetime. Only for higher He gas temperatures (above 10 K) the exciplex formation becomes again possible because the He atoms have enough kinetic energy to overcome the potential barrier.

When exciting the system at $R_b(5S)=6$ Å on the D_2 transition the corresponding fine-structure relaxation channel allows the system to form the terminal exciplex in the potential well of the $A^2\Pi_{1/2}$ state.

In solid helium only the largest exciplex $Rb^*He_{n_{max}}$ is observed after D_1 excitation. This means that the potential well is filled up to the maximal value of helium atoms that it can hold on a time scale which is shorter than the radiative lifetimes of the intermediate products. It is therefore likely to assume, as we have previously done for the formation of the corresponding cesium exciplexes that the exciplex results from a collective motion of the helium atoms. The difference to the experiments in gaseous He is that in those experiments at any temperature not only the terminal exciplex but also transient products were observed [2].

6.4.6 Summary and conclusion

We have presented several new spectral features observed in the laser-induced fluorescence from a helium crystal doped by laser ablation from a solid rubidium target. We detected for the first time weak, but unambiguously identified D_1 and D_2 fluorescence lines from atomic rubidium, which were previously believed to be completely quenched in solid helium. We have shown that Rb*He$_n$ exciplex formation is possible after D_1 excitation, in contrast to cesium doped solid He, in which exciplex formation proceeds mainly via absorption on the D_2 transition. We have explained this in terms of the smaller bubble diameter of rubidium, which allows the excitation to proceed directly to a binding state without tunnelling processes as they are needed with cesium. We have further reported the observation of Rb*He$_{1,2}$ exciplex emission after D_2 excitation, a process which could not be observed in liquid helium. Our study has clearly confirmed that there are marked differences between the spectroscopy of Rb in gaseous He and in solid He. From the point of view of model calculations liquid and solid He should behave in a similar way. We have also observed a larger exciplex. The main decay channel of laser excited Rb in solid helium is via the formation of this largest exciplex, assigned to be either Rb*He$_6$ or Rb*He$_7$ with subsequent emission of strongly red shifted fluorescence.

We proposed that the formation of a diatomic bubble could explain why we could observe the two exciplexes Rb*He$_1$ and Rb*He$_2$, while in equivalent experiment with cesium only the Cs*He$_2$ complex was detected. This feature could be related to a recently discovered dissociation band of the Rb$_2$ dimer which overlaps with the D_2 atomic absorption line [21]. This interpretation in terms of the diatomic bubble may also explain the absence of the Rb*He$_1$, Rb*He$_2$ and D_2 emission in liquid He. Because of the preparation process in liquid He, the Rb$_2$ dimer density may be strongly reduced. On the other hand it could also be, that due to the different pressure shifts the dissociation band of Rb$_2$ and the atomic absorption lines do not longer overlap in liquid He. More studies are needed to clarify this point. Besides purely spectroscopic studies, time-resolved femtosecond pump-probe experiments would be an additional helpful tool to elucidate this open question.

ACKNOWLEDGMENTS We thank J. Pascale for sending us his numerical Rb-He pair potentials. This work was supported by the grant number 200020-103864 of the Schweizerischer Nationalfonds.

References

[1] K. E. K. Hirano, M. Kumakura, Y. Takahashi, and T. Yabuzaki, Phys. Rev. A **66**, 042505 (2002).

[2] K. Hirano, K. Enomoto, M. Kumakura, Y. Takahashi, and T. Yabuzaki, Phys. Rev. A **68**, 012722 (2003).

[3] D. Nettels, A. Hofer, P. Moroshkin, R. Müller-Siebert, S.Ulzega, and A. Weis, Phys. Rev. Lett. **94**, 063001 (2005).

[4] J. Dupont-Roc, Z. Phys. B **98**, 383 (1995).

[5] S. Kanorsky, A. Weis, M. Arndt, R. Dziewior, and T. Hänsch, Z. Phys. B **98**, 371 (1995).

[6] O. Bünermann, M. Mudrich, M. Weidemüller, and F. Stienkemeier, J. Chem. Phys. **121**, 8880 (2004).

[7] C. P. Schulz, P. Claas, and F. Stienkemeier, Phys. Rev. Lett. **87**, 153401 (2001).

[8] F. R. Brühl, R. A. Trasca, and W. E. Ernst, J. Chem. Phys. **115**, 10220 (2001).

[9] J. Reho, J. Higgins, C. Callegari, K. K. Lehmann, and G. Scoles, J. Chem. Phys. **113**, 9686 (2000).

[10] J. Reho, J. Higgins, K. K. Lehmann, and G. Scoles, J. Chem. Phys. **113**, 9694 (2000).

[11] K. Enomoto, K. Hirano, M.Kumakura, Y. Takahashi, and T. Yabuzaki, Phys. Rev. A **69**, 012501 (2004).

[12] P. Moroshkin, A. Hofer, D. Nettels, S. Ulzega, and A. Weis, J. Chem. Phys. **124**, 024511 (2006).

[13] T. Kinoshita, K. Fukuda, T. Matsuura, and T. Yabuzaki, Phys. Rev. A **53**, 4054 (1996).

[14] T. Eichler, R. Müller-Siebert, D. Nettels, S. Kanorsky, and A. Weis, Phys. Rev. Lett. **88**, 123002 (2002).

[15] J. Pascale, Phys. Rev. A **28**, 632 (1983).
[16] R. A. Aziz and A. R. Janzen, Phys. Rev. Lett. **74**, 1586 (1995).
[17] S. I. Kanorsky, M. Arndt, R. Dziewior, A. Weis, and T. W. Hänsch, Phys. Rev. B **50**, 6296 (1994).
[18] T. Kinoshita, K. Fukuda, Y. Takahashi, and T. Yabuzaki, Phys. Rev. A **52**, 2707 (1995).
[19] H. Bauer, M. Beau, B. Friedl, C. Marchand, K. Miltner, and H. J. Reyher, Physics Letters A **146**, 134 (1990).
[20] T. Kinoshita, K. Fukuda, and T. Yabuzaki, Phys. Rev. B **54**, 6600 (1996).
[21] P. Moroshkin, A. Hofer, S. Ulzega, and A. Weis, arXiv:physics/0606100, to be published in Phys. Rev. A. (2006).
[22] P. Moroshkin, A. Hofer, S. Ulzega, and A. Weis, to be published (2006).

Part III

Magneto-optical experiments in solid ^{4}He

Chapter 7

Introduction to magneto-optical experiments in solid ^{4}He

7.1 Magnetic resonance experiments

Magnetic resonance experiments require a high degree of spin polarization of the atomic sample under investigation. The orientation changes of the macroscopic magnetization associated with the spin polarization of the sample, under exposure to a resonant oscillating magnetic field is the essence of magnetic resonance. Due to the quantum nature of solid He, it is a very well suited environment for high precision spin physics on paramagnetic impurity atoms. Implanted Cs atoms reside in spherical diamagnetic cavities (atomic bubbles) and have a very long relaxation time of spin polarization. Helium has neither a nuclear nor an electronic magnetic moment, so in first order it does not couple to the spin of the Cs valence electron. In conventional rare gas matrices, impurity atoms reside on lattice or defect sites and the spin polarization is rapidly destroyed due to the interaction with strong local crystal fields.

Conventional magnetic resonance experiments with particle densities on the order of 10^{18} cm^{-3} use high magnetic fields (several Tesla) and low temperatures to create a population difference between the spin states, compared to the Boltzmann distribution at zero field and at room temperature. The alteration of magnetization in conventional MR experiments is detected by pick-up coils. In our experiment we use a different technique. The Cs ground state consists of two hyperfine levels (F=3 and F=4) separated by 9.2 GHz. In thermal equilibrium the Zeeman sub-levels of the two hyperfine states are equally populated within each manifold and with the low atomic density (10^9 cm^{-3}) the conventional technique to detect alterations of the magnetization can not be used.

However the spin polarization can be created in a very efficient way by optical pumping, a technique invented by Brossel and Kastler [1] in 1949. They demonstrated that a population difference between the Zeeman sub-levels in the ground state can be created by irradiating paramagnetic atoms in a vapor cell with circularly polarized light (σ^+). The basic idea is to transfer angular momentum from the photon to the atoms. In this way the redistribution of populations in the ground state leads to a spin orientation.

Optical pumping of Cs atoms in solid ^{4}He was first observed by Weis et al. [2]. Later it was studied in detail by Lang et al. [3] and they demonstrated that the pumping mechanism of Cs in bcc solid He is of the Kastler-type. One speaks of Kastler-type pumping or repopulation pumping, when the spin polarization is preserved in the excited state, in contrast to the so-called depopulation pumping, where the spin polarization in the excited state is destroyed.

For optical pumping of Cs we use the D_1 ($6S_{1/2} \rightarrow 6P_{1/2}$) transition. The homogenous line width of the optical line in solid He is large ($\approx$ 10 nm) and therefore the hyperfine structure of the ground and the excited state can not be resolved optically. The large width makes also that the transition is not saturated when light intensities below 1 mW are used, so that stimulated emission processes can be neglected. The excited state decays only via spontaneous emission back to the ground state.

When the atomic sample is irradiated with resonant (D_1-transition) circularly polarized light each absorbed photon transfers an angular momentum of $+\hbar$ to the atom and drives therefore $\Delta M = +1$ transitions. The excited state decays through all allowed decay channels ($\Delta M = 0, \pm 1$) back to the ground state. On average the populations in the ground state are thus transferred to Zeeman levels with larger quantum numbers M. After a few absorption-emission cycles most of the population ends up in the level with quantum number $M = 4$, also called dark state. This state does not couple to the light which can be observed via a drop of the fluorescence light intensity once the dark state is populated. The optical pumping process is illustrated in Fig. 7.1

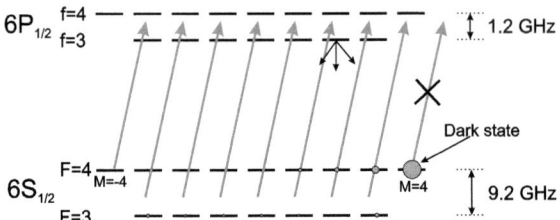

FIGURE 7.1: *Cs atom irradiated with resonant circularly polarized light. The hyperfine levels are not resolved due to the broad absorption line of Cs in solid He. After a few pumping cycles mainly the state with the highest magnetic quantum number $M = 4$ is populated. This state does not couple to the light and is called dark state.*

7.1.1 Optically detected magnetic resonance (ODMR)

The optical properties of an atomic sample change if the sample is spin polarized as explained in the previous section. Optically detected magnetic resonance is our standard technique to detect magnetic resonance transitions between Zeeman levels in the ground state by optical means. The standard way is to apply a static magnetic field B_0 parallel to the direction of the circularly polarized pumping laser beam (the so called M_z geometry). The populations in the ground state are transferred to the dark state to a large extend to the dark state when the sample is irradiated with circularly polarized resonant light (D_1 transition). The static magnetic field removes the degeneracy of the Zeeman sub-levels. We apply then an oscillating rf-field $B_1(t)$ with frequency ω_{rf} perpendicular to the static magnetic field. When the rf-frequency ω_{rf} matches the Larmor frequency $\omega_L = g\,\mu_B\,B_0/\hbar$, transitions between Zeeman sub-levels with different magnetic quantum number M are induced leading to a depopulation of the dark state, that was created by optical pumping. These transitions can be detected by an increase of the fluorescence light. The pumping laser is thus used to create the spin polarization and also to probe alterations of this polarization. Figure 7.2 shows a typical ODMR signal of Cs recorded in the bcc phase of solid ^{4}He in the M_z geometry.

We also performed ODMR experiments in the M_x geometry in which the system is operated as a phase-stabilized magnetometer. The two different geometries (M_z and M_x) are illustrated in Fig. 7.3. In the M_x geometry the static magnetic field is oriented in 45° with the direction of the laser beam. In this case the fluorescence light becomes modulated at the frequency ω_{rf}, which allows us to stabilize the frequency of the oscillating magnetic field $B_1(t)$ to the Larmor frequency ω_L. By measuring the frequency of the oscillating field, one can directly follow changes in the Larmor frequency without scanning across the resonance as in the M_z geometry. Both techniques were used for the Stark effect measurements.

7.2 Principle of EDM and Stark effect measurements

The measurement of an EDM was the original motivation for the matrix isolation experiments of alkali atoms in solid He, pioneered by A. Weis and S. Kanorsky in the early 1990's at the Max-Planck-Institute for Quantum Optics [4, 5]. This far-reaching goal of the experiment motivated our interest in theoretical and experimental investigations of the Stark effect, i.e., the interaction of the atom with an external static electric field. In this section we will discuss the general concept of an EDM measurement and we will explain why, after 10 years of efforts, the goal to measure an EDM had to be abandoned.

7.2 Principle of EDM and Stark effect measurements

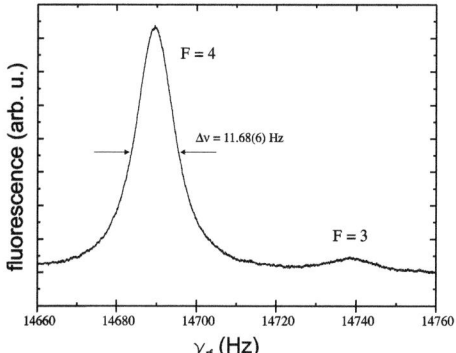

FIGURE 7.2: *Typical magnetic resonance curve of Cs recorded in a bcc crystal. The spectrum consists of two resonances arising from the two different hyperfine states of the $6S_{1/2}$ ground state, which have slightly different gyromagnetic ratios, and therefore different resonance frequencies.*

7.2.1 Stark effect

In our experiment with Cs atoms trapped in solid He, an EDM measurement would consist in measuring the shift of a magnetic resonance line (transition between different M-states) linear in the applied static electric field $\mathbb{E}$ by using the ODMR technique. The linear effect is superposed on a quadratic effect which is at least six orders of magnitude larger. The applied static electric field needed for an EDM measurement induces a dipole moment in the atom under investigation, which interacts with the external field and leads to a shift of the energy levels proportional to $\mathbb{E}^2$. This quadratic effect constitutes a large background for the EDM measurement and does not violate any discrete symmetry. So it is important to characterize this effect with high precision. This induced M-dependence of the effect can be parameterized in terms of an electric tensor polarizability $\alpha^{(3)}(F, M)$

$$\Delta E^{(3)} = -\frac{1}{2}\alpha^{(3)}(F,M)\mathbb{E}^2. \qquad (7.1)$$

The superscript (3) refers to the fact that the M-dependent effect arises only in third order perturbation theory. Details of the experiment and the theory will be given in chapters 8 and 9. We note here, that although this quadratic shift is large compared to the linear shift due to a possible EDM, it is seven orders of magnitude smaller than the overall shift of the ground state due to the second order polarizability $\alpha_0^{(2)}$.

A theoretical paper from the 1960's [6] predicted values for the electric tensor polarizability which were systematically larger than the experimental values for all five alkalis. The new theory developed by our group during the thesis work of Ulzega [7] brought the theoretical value for Cs to agree with the experimental one [8]. Details will be given in chapter 9. The measurements of the quadratic Stark effect in the Cs ground state in solid He has yielded a value for the electric tensor polarizability which differs slightly from the one of the free Cs atom [9, 10]. Details of this experiment are presented in chapter 8. The effect of the He matrix was taken into account in this work and results are presented in chapter 9. Experiment and theory are now in good agreement and show that the He matrix affects the electric tensor polarizability at a small level.

7.2.2 General concept of an EDM measurement

One of the basic principles in physics is the conservation of the combined three discrete symmetry operations **C** (charge conjugation), **P** (parity) and **T** (time reversal). The **CPT** theorem states that any

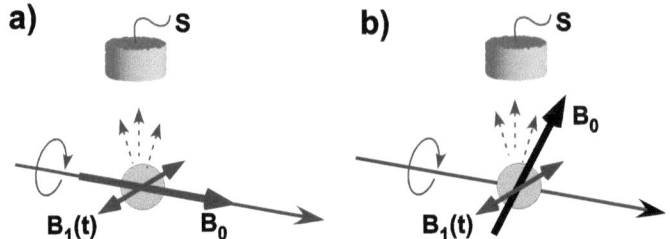

FIGURE 7.3: *Two different geometries for ODMR experiments with a static magnetic field B_0 and an oscillating field $B_1(t)$: (a) M_z geometry in which the frequency of the oscillating field is scanned across the Larmor frequency and the fluorescence shows a resonant change when the two frequencies coincide. (b) M_x geometrie; the fluorescence is modulated at the frequency of the oscillating field. The frequency can be actively stabilized to the Larmor frequency ω_L of the atoms. The static field B_0 is at 45° with respect to the laser beam.*

Lorentz invariant local quantum field theory with a Hermitian Hamiltonian must obey the **CPT** symmetry. The CPT theorem only demands the conservation of the combined symmetries, if one or two symmetries are violated it must be compensated by a violation of the third one.

Before the 1950's the general belief was that the mirror image of an elementary particle is the same as the particle itself (parity conservation). 1957 three independent experiments [11, 12, 13] have demonstrated the violation of parity in the weak decay of the ^{60}Co nuclei, of pions (π^+) and of muons (μ^+). These experiments formed the basis for a deeper understanding of the weak interactions and paved the way towards the Standard Model of particles. The violation of the discrete symmetry **P** is now well understood in the frame of the Standard Model. Parity violation in atoms was observed in the 1980's and refined since, yielding experimental results in agreement with the Standard Model predictions.

The violation of **T** or, in other words, the violation of the combined symmetry **CP** is still not completely understood. The first **CP** violation was observed in 1964 in decay channels of the neutral K-meson, an observation which led to the Nobel Prize for James Cronin and Val Fitch in 1980. The only direct observation of a **T**-violating effect was measured in 1998 in neutral Kaons [14].

The search for **T** violating effects is a way to look for physics beyond the Standard Model. S. Weinberg (Nobel Prize winner in physics 1979) made in 1992 the following statement:

> "...it may be that the next exciting thing to come along will be the discovery of a neutron or electron electric dipole moment.
>
> These electric dipole moments...seem to me to offer one of the most exciting possibilities for progress in particle physics."

An electric dipole moment of an elementary particle would violate time reversal symmetry. An elementary particle has an electric dipole moment if its center of mass and its center of charge do not coincide. The classical definition is

$$\vec{R}_{cg} = \int \vec{r} d^3 r \neq \int \rho_q \vec{r} d^3 r = \vec{d}_e. \tag{7.2}$$

For a point like particle the expression for the EDM can be found in [15].

The Standard Model allows elementary particles to have an extremely small electric dipole moment (electron EDM $|\vec{d}_e| < 10^{-37} e \cdot cm$) far beyond the sensitivity of present experiments. But other theories like SUSY or String Theory predict values that could be reached by next generation experiments. The experimental upper limit for the electron EDM is $|\vec{d}_e| < 1.6 \cdot 10^{-27} e \cdot cm$ [16]. If one blows an electron with a classical radius of 2.8 fm up to the size of the earth, this corresponds to a displacement of the center of mass with respect to the center of charge of only 35 nm.

The basic principle of an EDM measurement is to build an apparatus with a **P**-pseudoscalar property χ in order to detect a **P**- and **T**-pseudoscalar property of the atom [4, 5]. A possible apparatus pseudoscalar

7.2 Principle of EDM and Stark effect measurements

is the scalar product of a static magnetic and electric field $\chi = \vec{B} \cdot \vec{E}$. The magnetic field $\vec{B}$ is an axial vector which does change sign under the time reversal operation (and not under the parity operation **P**) whereas the electric field $\vec{E}$ does not change sign under **T** (but does under **P**-operation). One can define a pseudoscalar property of the atom χ_{atom} by the scalar product of the static electric dipole moment $\vec{d_e}$ and the magnetic moment $\vec{\mu} = g\,\mu_{bohr}\vec{S}/\hbar$ proportional to the spin $\vec{S}$, $\chi_{atom} = \vec{d_e} \cdot \vec{\mu}$. The Wigner-Eckart theorem implies the proportionality of the matrix elements for all vector quantities of a given system. In other words $\vec{d_e}$ has to be parallel or antiparallel to $\vec{\mu}$ and one can write $\vec{d_e} = d\,\vec{S}/\hbar$, where d is a proportionality factor. Figure 7.4 illustrates why an EDM violates the time reversal symmetry **T** and the parity operation **P**. The axial vector $\vec{\mu}$ changes sign under **T**-symmetry operation but the dipole moment $\vec{d_e}$ does not. If the time reversal symmetry is not violated then the particle (electron) and its time reversed image (**T**) have to occur in nature with equal probability. Thus the electric dipole moment $\vec{d_e}$ has to be parallel and antiparallel to the magnetic moment $\vec{\mu}$ at the same time, which is only possible if the electric dipole moment is zero. A non zero electric dipole moment violates both time reversal and parity symmetry. In an experiment one can mimic the time reversal operation by reversing the relative orientation of static electric and magnetic fields. The experimental signature of an EDM is a shift of a magnetic resonance line linear in the strength of an applied static electric field.

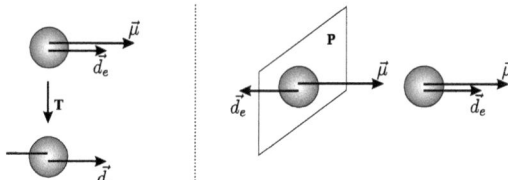

FIGURE 7.4: *Behavior of electric dipole moment and magnetic moment under the two discrete symmetry operations **T** and **P**.*

7.2.3 EDM measurement in solid He

A permanent electric dipole moment of the electron is enhanced in paramagnetic atoms. The enhancement factor R is proportional to the third power of the atomic charge Z

$$R = \frac{\text{EDM}_{atom}}{\text{EDM}_{electron}} \sim \alpha^2 Z^3. \tag{7.3}$$

The alkali atom Cs is therefore suited for an EDM measurement because it has a charge Z=133 and a simple electronic structure with only one valence electron which makes model calculations possible. Due to the long spin relaxation times and the high electric breakdown voltage of solid He it was thought that Cs atoms in solid He could be used to search for an EDM of the electron [4].

After a decade of experimental effort we came to the conclusion that our cryogenic system (Cs atoms in a bcc He crystal) cannot compete with present and ongoing experiments aiming an EDM measurement of a fundamental particle. This conclusion is based on several fundamental and technical limitations of the experiment that we have met in the past years. One of the main reasons is the limitation on the applicable static electric field to about 50 kV/cm. Higher fields result in electric breakdowns in the cell and a destruction of the sample. The breakdown voltage of a pure He crystal is larger than 50 kV/cm, but the ablated and implanted material, in particular unavoidable charged particles (electrons and charged clusters), reduces the electric strength that the sample can hold. Breakdowns outside of the cell occur at slightly higher fields but this limit could be pushed by developing better high voltage cables and feedthroughs. Another problem is the heat produced by the leakage current between the electrodes inside the cell (a few μA), which melts the crystal when a strong field is applied for more than one minute, and thereby destroys the sample. The magnetic fields associated with the leakage current constitute another serious systematic effect.

Although the signal to noise ratio of the magnetic resonance lines and their widths enable us to determine the position of the resonance with a precision below 100 mHz [17] in one second, sudden jumps of the line occur (probably related to the pulses of the Nd:YAG laser applied every 30 seconds to the sample in order to keep a reasonable atomic density), and limit this precision to 1 Hz. In addition to the sudden jumps we observe slow drifts (2 mHz/s) of the line, the origin of which is not yet clear [7]. During the ablation process atoms, clusters and also ions and electrons are produced. The charged particles can have an inhomogeneous distribution over the sample and thus produce different Stark shifts in the Cs atoms across the sample. The Nd:YAG laser pulses can lead to a sudden rearrangement of those charged particles or the Cs atoms and can produce sudden jumps of the zero field magnetic resonance line. Last, but not least, long data acquisition times are needed for a competitive EDM experiment. Many practical and technical difficulties limit the practical measuring time with a given He crystal to several hours only. One experiment typically lasts for two days (and two nights!), the time until the liquid He in the continuously pumped helium bath is evaporated. Out of these two days approximately 10 hours can be used for meaningful data taking, the rest of the time is needed for growing He crystals (4-5 times per experiment) and aligning the excitation laser and the detection system in order to maximize the spectroscopic signals. For a competitive EDM experiment several 100 hours of continuous data taking are required, not possible with the present system.

References

[1] J. Brossel and A. Kastler, Comptes rendu hebdomadaires des seances de l'academie des sciences **229**, 1213 (1949).

[2] A. Weis, S. Kanorsky, M. Arndt, and T. W. Hänsch, Z. Phys. B **98**, 359 (1995).

[3] S. Lang, S. I. Kanorsky, T. Eichler, R. Müller-Siebert, T. W. Hänsch, and A. Weis, Phys. Rev. A **60**, 3867 (1999).

[4] M. Arndt, S. I. Kanorsky, A. Weis, and T. W. Hänsch, Phys. Lett. A **174**, 298 (1993).

[5] A. Weis, S. Kanorsky, S. Lang, and T. W. Hänsch, in *Lecture Notes in Physics* (Springer, 1997).

[6] P. G. H. Sandars, Proc. Phys. Soc. **92**, 857 (1967).

[7] S. Ulzega, Ph.D. thesis, University of Fribourg, Switzerland (2006).

[8] S. Ulzega, A. Hofer, P. Moroshkin, and A. Weis, Europhys. Lett. **76**, 1074 (2006).

[9] C. Ospelkaus, U. Rasbach, and A. Weis, Phys. Rev. A **67**, 011402(R) (2003).

[10] S. Ulzega, A. Hofer, P. Moroshkin, R. Müller-Siebert, D. Nettels, and A. Weis, Phys. Rev. A **75**, 042505 (2007).

[11] C. S. Wu, E. Ambler, R. W. Hayward, D. D. Hoppes, and R. P. Hudson, Phys. Rev. p. 1413 (1957).

[12] R. Garwin, L. Lederman, and M. Weinrich, Phys. Rev. p. 1415 (1957).

[13] J. Friedman and V. L. Telegdi, Phys. Rev. p. 1681 (1957).

[14] A. Angelopoulos, A. Apostolakis, and E. Aslanides et al., Phys. Lett. **444**, 43 (1998).

[15] W. Bernreuther and M. Suzuki, Rev. Mod. Phys. **63**, 313 (1991).

[16] B. C. Regan, E. D. Commins, C. J. Schmidt, and D. DeMille, Phys. Rev. Lett. **88**, 071805 (2002).

[17] S. I. Kanorsky, S. Lang, S. Lücke, S. B. Ross, T. W. Hänsch, and A. Weis, Phy. Rev. A **54**, R1010 (1996).

Chapter 8

Paper V:
Measurement of the forbidden electric tensor polarizability of Cs atoms trapped in solid ^{4}He

This paper reports on an optically detected magnetic resonance (ODMR) experiment of Cs atoms implanted in a bcc ^{4}He crystal. The technique was used to detect shifts of the magnetic sublevels in the Cs ground state induced by an external static electric field. From the shifts we infer the tensor polarizability of Cs atoms.

My main contributions to the work were:

- Setting up the detection system for the two different configurations (M_x and M_z) for magnetic resonance experiments.

- Setting up the high voltage power supplies including the high voltage cables, that enter the cryostat. Developing software to control the two high voltage power supplies which allows fast switching of the electric field (positive, off, negative) and a more accurate control of the applied field.

- Determination of the separation of the glass electrodes in the pressure cell during the experiment at 1.5 K. The polycarbonate body that holds the electrodes shrinks considerably when going from room temperature to 1.5 K.

- Recording data together with P. Moroshkin and S. Ulzega.

- Producing figures and text for the paper.

Measurement of the forbidden electric tensor polarizability of Cs atoms trapped in solid ^{4}He

S. Ulzega[1], A. Hofer[2], P. Moroshkin[2], R. Müller-Siebert[3], D. Nettels[4] and A. Weis[2]

[1] *EPFL, Lausanne, Switzerland. Email address: simone.ulzega@epfl.ch*
[2] *Département de Physique, Université de Fribourg, Chemin du Musée 3, 1700 Fribourg, Switzerland*
[3] *SELFRAG AG, Langenthal, Switzerland*
[4] *Biochemisches Institut, Universität Zürich, Switzerland*

Published in Phys. Rev. A **75**, 042505 (2007).

Abstract: We have measured the electric tensor polarizabilities of the hyperfine levels of Cs atoms embedded in a body-centered cubic (bcc) ^{4}He crystal. The polarizabilities are inferred from the shift of optically detected magnetic resonance lines in each hyperfine level induced by static electric fields up to 50 kV/cm. We recorded the magnetic resonances both by scanning the rf frequency and in a configuration in which the system is operated as a phase-stabilized magnetometer. The results from both measurements agree well with model calculations taking the effect of the solid helium matrix and our recent extension of the theory of forbidden tensor polarizabilities into account. We have also performed the first measurement of the differential tensor Stark splittings of the F=3 and F=4 hyperfine levels of the ground state, thus confirming the recently revised sign of this shift which affects the blackbody correction of primary frequency standards.

8.1 Introduction

The Stark effect, i.e., the effect of a static electric field on atomic properties is one of the fundamental interactions in atomic physics. In this paper we address tiny modifications of the energy of the magnetic sublevels of the cesium ground state induced by the tensor part of the electric interaction.

The Stark effect of the atomic hyperfine structure was treated in a comprehensive paper by Angel and Sandars [1], who showed that in second order perturbation theory the Stark shift of a level $|\gamma\rangle = |nL_J, F, M\rangle$ can be parametrized in terms of scalar, $\alpha_0^{(2)}$, and tensor, $\alpha_2^{(2)}$, polarizabilities. As tensor polarizabilities have non-zero values for states with $L \geq 1$ only, the spherically symmetric $nS_{1/2}$ ground state of alkali atoms has only a scalar polarizability and all its magnetic sub-levels $|F, M\rangle$ are expected to experience the same Stark shift, independent of F and M. However, it has been experimentally known since several decades that an electric field leads to F-dependent [2] and M-dependent [3] energy shifts in the alkali ground states. The latter effect is described by a (forbidden) tensor polarizability α_2. Improved measurements of the ground state tensor polarizabilities were performed by Carrico et al. [4] and Gould et al. [5] using conventional atomic beam Ramsey resonance spectroscopy. A recent remeasurement of the tensor polarizability of ^{133}Cs in an all optical atomic beam experiment [6] has confirmed the earlier experimental values [4, 5] of $\alpha_2(F = 4)$ of cesium.

In 1967 Sandars [7] showed that the F- and M-dependence of the Stark effect can be explained by extending the perturbation theory to third order after including the hyperfine interaction. The theoretical expression for the tensor polarizability $\alpha_2^{(3)}(F = 4)$ given in [7] was evaluated numerically in [3] and [5] under simplifying assumptions. The comparison with the experimental polarizabilities showed that the absolute theoretical values were systematically larger for all five alkalis studied in [5]. We note that in cesium the third order Stark splittings of the Zeeman levels due to $\alpha_2^{(3)}$ are approximately 7 orders of magnitude smaller than the overall shift of the ground state due to the second order polarizability $\alpha_0^{(2)}$.

In a recent paper [8] we have revisited the third order Stark theory by identifying and evaluating contributions which were not included in the earlier calculations. This led to a theoretical value of the tensor polarizability of Cs which is in good agreement with all existing experimental results [4, 5, 6]. As described in [8] we have also identified a sign error in the previous treatment [7] of the Stark effect concerning the relative signs of the tensor polarizabilities of the two ground state hyperfine levels. We have shown that this relative sign has a direct implication for the precise evaluation of the blackbody radiation shift of the hyperfine transition frequency from static Stark shift measurements [8].

The initial motivation for our experimental [6] and theoretical [8] studies of the third order Stark interaction was the long standing discrepancy between experimental and theoretical values of $\alpha_2^{(3)}$. With our revised theory [8] this 40-year-old problem has found a satisfactory solution. In the work reported here we have extended the study of strongly suppressed Stark interactions to cesium atoms trapped in a solid ^{4}He matrix. We have measured (and calculated) the tensor polarizability of Cs in a quantum solid matrix using two independent experimental techniques. Both methods yield consistent values of the tensor polarizability $\alpha_2^{(3)}$ whose moduli are approximately 10% larger than $|\alpha_2^{(3)}|$ of the free cesium atom. A calculation [9] which considers the influence of the helium matrix on the atomic energies and wave functions entering the third order perturbation theory can account for this matrix-induced shift. In addition we have made the first experimental determination of the relative sign and magnitude of the tensor polarizabilities of the two ground state hyperfine levels and thereby confirmed the sign predicted by our model calculations.

The extension of the Stark effect investigations to solid helium was motivated by our past proposal [10, 11] that alkali atoms in condensed helium matrices might be an interesting sample to search for a permanent electric dipole moment (EDM) of the electron. In such experiments the quadratic Stark shift constitutes a strong background and its (imperfect) suppression is a major source of systematic uncertainty. While the perturbation of optical and magnetic properties of alkali atoms by condensed (superfluid and solid) ^{4}He matrices has been studied extensively in the past decade [12] the effect of the solid He environment on static electric properties has never been addressed. To our knowledge the present study is the first investigation of the Stark effect of atomic defects in condensed helium. The study confirms that the so-called extended bubble model is well suited for the quantitative description of such matrix-induced perturbations of electric polarizabilities.

8.2 Theory

The perturbation of the energies of ground state magnetic sublevels $|\gamma\rangle = |6S_{1/2}, F, M\rangle$ by a static electric field $\mathbb{E}$ is conventionally parametrized in terms of an electric polarizability $\alpha(\gamma)$ according to

$$\Delta E(\gamma) = -\frac{1}{2}\alpha(\gamma)\,\mathbb{E}^2\,. \qquad (8.1)$$

The polarizability $\alpha(\gamma) = \alpha^{(2)}(\gamma) + \alpha^{(3)}(\gamma)$ has contributions from perturbations arising in second and third order perturbation theory. The second order polarizability $\alpha^{(2)}$ gives rise to the well studied scalar, i. e., F- and M- independent shift of the levels. The total third order polarizability can be written as [8]

$$\alpha^{(3)}(6S_{1/2}, F, M) =$$
$$\alpha_0^{(3)}(6S_{1/2}, F) + \alpha_2^{(3)}(6S_{1/2}, F)\frac{3M^2 - F(F+1)}{2I(2I+1)}f(\theta),$$

$$(8.2)$$

where the function $f(\theta) = 3\cos^2\theta - 1$ describes the orientation of the electric field with respect to the quantization axis, and where I is the nuclear spin. The first term of Eq. (8.2), $\alpha_0^{(3)}$, is an F-dependent scalar contribution to the third order polarizability. It gives the main contribution to the Stark shift of the hyperfine transition frequency (F-dependent effect) but does not alter the Zeeman sub-structure of the ground state. The second term, described by the third order tensor polarizability $\alpha_2^{(3)}$, produces F- and M^2-dependent energy shifts. Its main effect is the removal of the Zeeman degeneracies between the magnetic sublevels within each of the two ground state hyperfine levels (M^2-dependent effect). The tensor term, evaluated for M=0, also gives an additional small contribution ($\approx 1\%$) to the shift of the hyperfine transition frequency. A straightforward way to measure the effect of the tensor polarizability is the observation of an electric field induced shift of magnetic resonance transition frequencies within a given hyperfine multiplet. Because of the selection rules magnetic resonances can only be driven between adjacent magnetic sublevels $|F, M\rangle \to |F, M - 1\rangle$. For Cs (I=7/2) the differential Stark shift of that transition can be calculated from Eqs. (8.1) and (8.2) to be

$$\Delta\nu_{(|F,M\rangle \to |F,M-1\rangle)} = -\frac{3}{56}\frac{\alpha_2^{(3)}(F)}{h}(2M-1)\,\mathbb{E}^2\,.$$
$$(8.3)$$

The third order polarizabilities involve both the dipole-dipole and the electric quadrupole hyperfine interactions, so that $\alpha_2^{(3)}$ of the two hyperfine levels $F = 3, 4$ can be expressed [8] in terms of these contributions as

$$\alpha_2^{(3)}(F=4) = a_1 + a_2\,, \qquad (8.4a)$$
$$\alpha_2^{(3)}(F=3) = -a_1 + \frac{5}{3}a_2\,, \qquad (8.4b)$$

where a_1 and a_2 are due to the dipole-dipole and to the quadrupole interaction, respectively. The latter contribution is very small ($a_2/a_1 \approx 4 \cdot 10^{-4}$) and can be neglected. With this approximation the tensor polarizabilities of the two ground state hyperfine levels are thus connected by the simple relation $\alpha_2^{(3)}(3) \approx -\alpha_2^{(3)}(4)$. This result is in contradiction with an earlier work [7] which predicts the same sign for the tensor polarizabilities of the two ground state hyperfine levels. All measurements of tensor polarizabilities published to date were performed in the $F = I + 1/2$ hyperfine states, so that no prior experiment was sensitive to the relative signs of $\alpha_2^{(3)}(3)$ and $\alpha_2^{(3)}(4)$. Below we will present experimental evidence for the correctness of the sign derived in our calculation.

8.3 Experimental methods

8.3.1 Helium matrix isolation spectroscopy

Alkali atoms embedded in the isotropic body-centered cubic (bcc) phase of ^{4}He impose their symmetry on the local matrix environment thereby forming spherically symmetric cavities (atomic bubbles). The isotropy of the trapping sites, together with the diamagnetic nature of the matrix lead to longitudinal spin relaxation times T_1 of 1-2 s [13]. This allows the efficient optical pumping of the sample and the observation of magnetic resonance linewidths below 20 Hz in optical-rf double resonance experiments [14]. Moreover, condensed helium has an electric break-down voltage in excess of 100 kV/cm, which makes it, in principle, an ideal environment for high resolution magnetic resonance experiments in strong electric fields.

8.3.2 The sample cell

The measurements reported below were performed on cesium atoms implanted in a solid ^{4}He matrix. The experimental setup is similar to the one described in [15]. The helium crystal is grown at

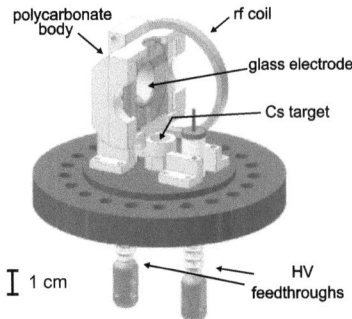

FIGURE 8.1: (Color online) The bottom flange of the pressure cell with one of the two rf-field coils and one of the two HV electrodes shown.

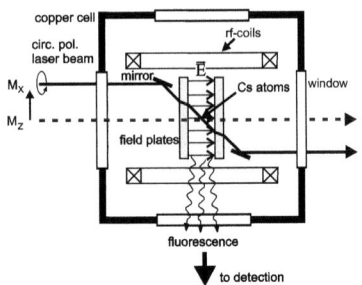

FIGURE 8.2: Top view of the set-up for magnetic resonance experiments with electric fields. The presence of two mirrors in the pressure cell allows us to switch between the M_x and M_z configurations with a simple translation of the laser beam.

pressures around 30 bar in a cubic copper pressure cell (inner volume=175 cm^3) immersed in superfluid helium cooled to 1.5 Kelvin by pumping on the helium bath. Optical access to the inner cell volume is given by four lateral windows and a top window. Laser excitation and fluorescence detection of the atoms occur through the side windows, while the top window is used for the implantation process. The host matrix is doped with cesium atoms by means of laser ablation with a pulsed, frequency-doubled Nd:YAG laser beam (2 Hz repetition rate) focused onto a solid Cs target located at the bottom of the cell. Diffusion of the implanted atoms and the subsequent binding into dimers and clusters leads to a drop of the atomic fluorescence signal with time. Low-energy pulses of the same Nd:YAG laser at a lower repetition rate are therefore used, once the crystal is doped, to dissociate dimers and clusters. In this way the average atomic density can be kept at a level of 10^8–10^9 cm^{-3}.

The pressure cell is surrounded by three orthogonal pairs of superconducting Helmholtz coils for applying a static magnetic field and for compensating residual fields, while another pair of Helmholtz coils mounted inside of the cell allows the application of an oscillating rf field for driving the magnetic resonance transitions (8.2). The cryostat is shielded from laboratory fields by a three-layer μ-metal shield.

The inner part of the cell, shown in Fig. 8.1, contains a split polycarbonate body which holds the rf-coils as well as two transparent glass electrodes which allow the application of the static electric field for the Stark effect experiments. The electrodes are (4 mm thick) quadratic float glass plates of 40×40 mm^2 whose facing surfaces are coated with a conductive tin oxide layer. Their opposite surfaces are partially coated with gold and electrically connected to the front surface by a vapor deposited gold stripe. Copper rings connected directly to low-temperature compatible HV feedthroughs containing no magnetic components are mechanically pressed onto the plates' back surfaces. The use of two feedthroughs allows us to charge each plate individually. The plate spacing of d=6 mm at room temperature expands to d=6.35(5) mm when the cell is cooled to 1.5 K. With the given surface/spacing ratio the field in the center deviates by much less than 1% from V/d.

The high voltage was generated by two identical power supplies of opposite polarities and delivered to the feedthroughs via HV cables traversing the top flange of the cryostat and the helium bath. In this way we were able to apply electric fields up to 50 kV/cm. This upper limit was due to sparking which occured both inside and outside of the pressure cell. The doping of the crystal by laser ablation produces atomic ions and charged clusters that lead to a leakage current of a few μA (at 25 kV) between the electrodes which locally melts the crystal and limits the maximum useful voltage.

A top view of the pressure cell for magnetic resonance experiments with electric fields is shown in Fig. 8.2. The use of two suitably oriented mirrors allowed the easy switching between the M_z configuration ($\widehat{k} = \widehat{\mathbb{E}}$) and the M_x configuration ($\widehat{k} \cdot \widehat{\mathbb{E}} = 1/\sqrt{2}$) described below by a simple horizontal translation of the laser beam.

8.3.3 The magnetic resonance technique

Optically detected magnetic resonance (ODMR) combines magnetic resonance with optical preparation and detection. It is a powerful method for performing magnetic resonance spectroscopy in dilute samples of paramagnetic atoms. A high degree of spin polarization is an essential prerequisite for observing magnetic resonance. In our experiments spin polarized cesium atoms are prepared by optical pumping [15] with circularly polarized laser light tuned to the D_1 line ($6S_{1/2} \rightarrow 6P_{1/2}$ transition). Due to the large homogeneous linewidth of the optical absorption line of Cs in condensed helium [12], the hyperfine structure of the transition is not resolved. After a number of absorption-emission cycles the majority of the atoms is pumped into the state $|F=4, M=4\rangle$ which does not absorb circularly polarized light. The polarized sample thus does not fluoresce and the sample is said to be in a *dark state*. Any subsequent depolarizing interaction, such as a magnetic resonance transition, leads to an increase of the fluorescence rate. This constitutes the basis of the optical detection of the magnetic resonance. Efficient optical pumping of alkali atoms embedded in the isotropic bcc phase of solid ^{4}He was demonstrated by Lang et al. [15].

The optical properties of the polarized sample depend on the orientation of the spin polarization with respect to the light beam. The magnetic resonance is driven by a weak oscillating magnetic field (called rf field below) applied perpendicularly to the main static field . When the oscillation frequency coincides with the Larmor precession frequency depolarizing transitions between adjacent sublevels are induced. This leads to a resonant change in the fluorescence rate when the rf frequency is tuned across the Larmor frequency. In practice the same laser beam that produces the spin polarization also detects its alteration by the magnetic resonance process. The laser-induced atomic fluorescence is imaged onto an avalanche photodiode, whose photocurrent is recorded by a digital oscilloscope. Background radiation from scattered laser light is suppressed by an interference filter.

8.4 Measurements

8.4.1 The tensor polarizability in the M_z geometry

The M_z geometry is characterized by the static magnetic field $\vec{B}_0$ being oriented parallel to the

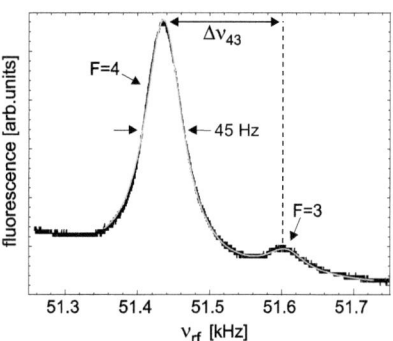

FIGURE 8.3: Magnetic resonances of the Zeeman transitions within the F=4 (left) and F=3 (right) hyperfine multiplets recorded with $\mathbb{E} = 0$ in the M_z geometry.

propagation direction $\widehat{k}$ of the pumping light and hence to the initially created spin polarization. In this case the magnetic resonance manifests itself as a resonant change of the DC level of the fluorescence signal.

The optical pumping process produces population imbalances between the magnetic sublevels in both hyperfine levels of the ground state. Because of the finite nuclear magnetic moment, the gyromagnetic ratios $\gamma(F)$ of these two states differ slightly, besides having opposite signs. As a consequence, the magnetic resonance transitions in the F=3 and F=4 states occur at slightly different frequencies and can be resolved in a single scan of the rf-field, as shown in Fig. 8.3. In the low magnetic fields used here the Zeeman effect is linear and all individual resonances in a multiplet of given F occur at the same frequency. The dominating components in the two lines of Fig. 8.3 correspond to the transitions $|4,4\rangle \rightarrow |4,3\rangle$ and $|3,3\rangle \rightarrow |3,2\rangle$. The spectrum is fitted by two Lorentzian lines superposed on a curved background $bg(\nu)$. This background is due to the slow disappearance of the atomic signal as atoms recombine into dimers and clusters. It was recorded in separate runs with no applied rf field and it is well fitted by the empirical function $bg(\nu) = b_1 \exp[-\lambda_1 \nu] + b_2 \exp[-\lambda_2 \nu]$. When an electric field is applied the magnetic resonance lines are displaced due to the M^2-dependent (differential) Stark shift of the sublevels coupled by the rf transition. This shift is proportional to $\alpha_2^{(3)} \mathbb{E}^2$. At each value of the electric field we have recorded spectra with each field polarity, in between which a spec-

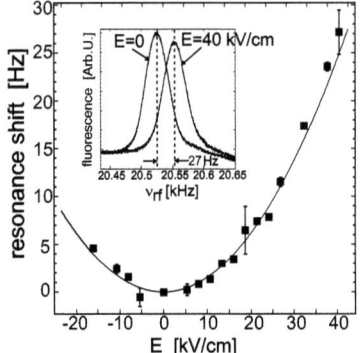

FIGURE 8.4: Stark shift of the Cs ground state magnetic resonance $|4,4\rangle \to |4,3\rangle$ in solid ^{4}He measured in the M_z geometry. Some points have an error bar smaller than the symbol size. The fit function is of the type $\Delta\nu_4 = \eta_z \mathbb{E}^2$. The different error bars are explained in Section 8.4.3. The inset shows magnetic resonance lines measured with and without applied electric field used to infer the Stark shift.

trum with no applied field was recorded. The latter reference measurements were necessary as we observed a slow drift of the zero electric field magnetic resonance frequency (cf Sect. 8.4.3), which represented the main limitation of the sensitivity of our apparatus.

The dependence of the line center of the $F = 4$ resonance on the electric field strength is shown in Fig. 8.4. According to Eq. 8.3 the Stark shift of the line is given by

$$\Delta\nu_4 = \eta_z \mathbb{E}^2 = -\frac{3}{8}\alpha_2^{(3)}(4)\,\mathbb{E}^2\,, \qquad (8.5)$$

under the assumption that the line consists only of the $|4,4\rangle \to |4,3\rangle$ transition. A quadratic fit to the data, shown as solid line in Fig. 8.4, then yields the tensor polarizability

$$\alpha_2^{(3)}(4) = -4.07(20) \times 10^{-2} \frac{Hz}{(kV/cm)^2}\,. \qquad (8.6)$$

This value is shown as point (d) in Fig. 8.11. The figure also shows previous experimental values of $\alpha_2^{(3)}(4)$ obtained on free Cs atoms in atomic beam experiments together with the corresponding theoretical value [8]. Eq. 8.6 assumes that the F=4 line consists only of the $|4,4\rangle \to |4,3\rangle$ transition, i.e., a 100% polarized sample. Because of the finite degree of spin polarization the recorded line contains a small admixture of the $|4,3\rangle \to |4,2\rangle$ transition. Based on the rate equation calculations described in Sect. 8.5 we find that this effect leads to an underestimation of $\alpha_2^{(3)}$ by less than 1%.

We have recently extended our calculations of the Cs tensor polarizability of free Cs atoms [8] to include the effect of the helium matrix [9]. This effect increases $\alpha_2^{(3)}$ by approximately 10 %, as shown in Fig. 8.11. The experimental result (Eq. 8.6) of the measurement in the M_z geometry is in good agreement with that theoretical calculation.

8.4.2 The relative sign of $\alpha_2(F = 4)$ and $\alpha_2(F = 3)$

In order to determine the relative sign of the polarizabilities in the F=3 and F=4 states we measured the splitting $\Delta\nu_{43}$ (introduced in Fig. 8.3) of the corresponding resonance frequencies, defined by $\Delta\nu_{43} \equiv \Delta\nu_{|4,4\rangle \to |4,3\rangle} - \Delta\nu_{|3,3\rangle \to |3,2\rangle}$. For this measurement we use the fact that both resonances can be observed in a single scan (cf. Fig. 8.3). This reduces the measurement time and thus systematic effects due, e.g., to line drifts as discussed below. The line centers are inferred from Lorentzian line fits. Fig. 8.5 shows the electric field dependence of the splitting $\Delta\nu_{43}$ between the resonances in F=3 and F=4. If one assumes $\alpha_2^{(3)}(4) = \alpha_2^{(3)}(3)$, as given in Sandars' work [7], one expects the dependence

$$\Delta\nu_{43} = -\frac{18}{28}\alpha_2^{(3)}(4)\mathbb{E}^2\,, \qquad (8.7)$$

shown as dotted line in Fig. 8.5. On the other hand, our recent calculation [8] predicts $\alpha_2^{(3)}(4) = -\alpha_2^{(3)}(3)$, which yields

$$\Delta\nu_{43} = -\frac{3}{28}\alpha_2^{(3)}(4)\mathbb{E}^2\,, \qquad (8.8)$$

a dependence shown as solid line in Fig. 8.5. The good agreement of the experimental data with the latter dependence proves that the tensor polarizabilities of the two hyperfine levels have indeed opposite signs as predicted by our calculation.

8.4.3 Line drifts

The sensitivity of the M_z configuration is limited by small instabilities of the magnetic resonance frequency (with and without applied electric field). We have made long-time recordings of the magnetic resonance signals under identical conditions and we can distinguish two distinct effects. First, on a scale of 45 minutes the resonance frequency shows a slow drift at a rate of 1.5–2.5 mHz/s which

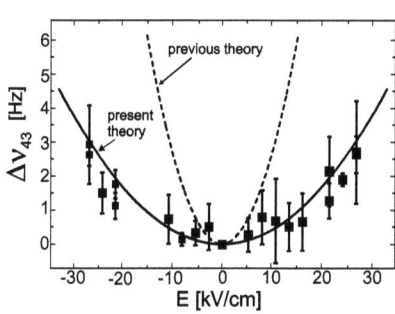

FIGURE 8.5: The differential shift $\Delta\nu_{43} \equiv \Delta\nu_{|4,4\rangle \to |4,3\rangle} - \Delta\nu_{|3,3\rangle \to |3,2\rangle}$. The dotted line represents the prediction of Sandars' work while the solid line is the prediction of our calculations, as explained in the text.

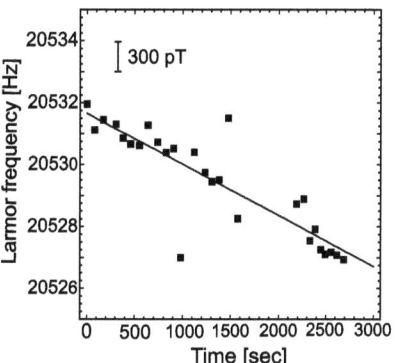

FIGURE 8.6: Zero-electric-field resonance drift. The solid line is a linear fit which gives a drift rate of $1.7\,\mathrm{mHz/s}$ corresponding to a magnetic field drift of $\approx 500\,\mathrm{fT/s}$.

is equivalent to a magnetic field drift rate of about $500\,\mathrm{fT/s}$. This frequency drift may also be associated with a slow motion of the center of gravity of the atomic sample due to atoms drifting in the He crystal, in combination with a magnetic field gradient. Our field has indeed a small gradient of $\approx 3\,\mathrm{nT/mm}$ [16], corresponding to a relative inhomogeneity of $2\cdot 10^{-4}/\mathrm{mm}$. In this case the resonance drift could be explained by an atomic drift velocity of $0.2\,\mu\mathrm{m/s}$.

A second effect occurs on a much shorter time scale. The Nd:YAG pulses, sent into the crystal every 30 seconds between sweeps of the rf frequency in order to dissociate clusters and to recover the optical fluorescence signal, can locally melt the crystal and occasionally provoke sudden drifts of the atoms which appear as steep jumps of their Larmor frequency (shown in Fig. 8.6). The error bars of Figs. 8.4 and 8.5 are related to such jumps. We determined the average of the line positions in zero field, measured before and after the actual Stark shift measurement. This average position was then subtracted from the line position measured with the field applied. In this way we could infer the Stark shift corrected for linear drifts of the base line. The error bar reflects the drift-induced variations of the two reference measurements. The relative importance of both effects was found to vary substantially from crystal to crystal, or in a given crystal after different atomic implantations. In order to reduce such effects we have performed a second series of measurements using an alternative magnetic resonance technique described in the next paragraph.

8.4.4 The tensor polarizability in the M_x geometry

The M_x geometry is characterized by the static magnetic field $\vec{B}_0$ being oriented at an angle β with respect to $\hat{k}$. In this case the magnetic resonance leads to a modulation of the transmitted laser power at the rf frequency with an amplitude varying as $\sin 2\beta$. The largest signal is thus obtained for $\beta = \pi/4$ (Fig. 8.2). In practice the modulation amplitude, which is resonant when the rf frequency matches the Larmor frequency is measured using a phase sensitive (lock-in) amplifier referenced to the rf frequency [14]. The in-phase and quadrature components of the demodulated signals have absorptive and dispersive Lorentzian line shapes and the phase φ of the signal modulation is given by

$$\tan\varphi = \frac{\delta\nu}{\Delta\nu}, \qquad (8.9)$$

where $\delta\nu = \nu_{rf} - \nu_L$ is the detuning of the rf frequency ν_{rf} with respect to the Larmor frequency ν_L, and where $\Delta\nu$ is the linewidth. The linear dependence of the phase signal $\varphi(\nu_{rf})$ near resonance ($\delta\nu \approx 0$) was used in a feed-back system generating the rf frequency: the phase signal, amplified by a PID controller, was used to drive a voltage controlled oscillator which generated the oscillating voltage driving the rf coils. In this way the rf frequency was phase-locked to the Larmor frequency. Any changes of the resonance condition, induced,

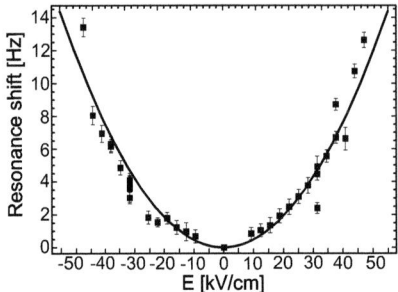

FIGURE 8.7: Stark shift of the Larmor frequency in the M_x configuration, measured using the M_x (self-oscillating) magnetometer configuration discussed in the text. The solid line is a fit function of the type $\Delta \nu_4 = \eta_x \, \mathbb{E}^2$.

e.g., by a drift of the magnetic field or by a displacement of the resonance frequency induced through the Stark effect can then be detected by a real-time monitoring of ν_{rf} with a frequency counter.

This operation of the system as a phase-locked magnetometer [17] has allowed us a faster measurement of the electric field induced changes of the Larmor frequency. Compared to the experiments in the M_z geometry it has the advantage of being less sensitive to systematic effects coming from slow drifts of system parameters. It has the drawback that the resonance frequencies in the $F = 3$ and $F = 4$ states cannot be measured simultaneously. We have used this method to record the quadratic electric field dependence of the resonance frequency in the $F = 4$ state. The results are shown in Fig. 8.7. A quadratic fit of the type $\Delta\nu_4 = \eta_x \mathbb{E}^2$ yields

$$\eta_x = 0.469(30) \times 10^{-2} \,\mathrm{Hz/(kV/cm)}^2 \,. \qquad (8.10)$$

In the M_z geometry, in which the pumping direction is along the magnetic field (quantization axis) the field stabilizes the polarization created by optical pumping and for a 100% polarized sample the tensor polarizability $\alpha_2^{(3)}$ is related to η_z by Eq. 8.5. In the M_x geometry, on the other hand, for which the pumping direction and the magnetic field direction, i.e., the axis of quantization are no longer parallel this simple relation no longer holds. As a consequence oscillating steady state populations appear in all 2F+1 $|F = 4, M\rangle$ states. This is illustrated in Figs. 8.8a,b where we compare the steady state populations produced by optical pumping of the sublevels in the $F = 4$ manifold in the M_z and in the M_x geometries respectively. These results were

obtained from a rate equation calculation described earlier [15]. The parameters of that calculation are the optical pumping rate, γ_p, proportional to the laser intensity, and the longitudinal spin relaxation rate, γ_1.

8.5 Analysis of the M_x data

Because of the tilted quantization axis the extraction of the tensor polarizability $\alpha_2^{(3)}$ from the data of Fig. 8.7 recorded in the M_x geometry requires a more detailed analysis. We base this analysis on the three step approach, discussed in [18], which is well suited for the quantitative description of optically detected magnetic resonance signals. In that model the double resonance process is treated as three time sequential processes, viz., the creation of steady state spin orientation by optical pumping, the evolution of that initial orientation under the influence of the external fields, and finally the optical detection of the steady state oscillation reached in the second step. The validity of this approach is discussed in the quoted reference.

The optical pumping process (step 1)

In the first step spin polarization (orientation) is created in the sample by optical pumping with circularly polarized resonance radiation. The interaction with the magnetic field and relaxation processes then lead to a steady state redistribution of the sublevel populations p_M, as shown in Fig. 8.8 for a given set of the parameters γ_p and γ_1. For the calculation one sets the quantization axis along the static field $\vec{B}_0$, and the populations produced in the first step are expressed in this coordinate system.

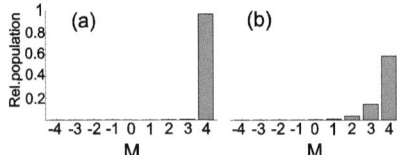

FIGURE 8.8: Steady state populations after optical pumping in the ground state level F=4 in the M_z configuration (a) and in the M_x configuration (b). The parameters (defined in [15]) of the calculation are the pumping rate $\gamma_p = 2500\,\mathrm{s}^{-1}$ and the longitudinal spin relaxation time $\gamma_1 = 1\,\mathrm{s}^{-1}$.

The magnetic resonance process (step 2)

In the magnetic resonance process the initial spin orientation evolves to a steady state precession under the joint action of the external fields $\vec{B}_0$ and $\vec{B}_1 \cos \omega_{rf} t$ and of relaxation. This evolution is described by the Liouville equation for the density matrix ρ

$$\dot{\rho} = -\frac{i}{\hbar}[H(t), \rho] + H_{\text{relax.}} . \qquad (8.11)$$

After applying the rotating wave (rw) approximation (coordinate system rotating at the frequency ω_{rf} around $\hat{B}_0$) the Hamiltonian becomes time independent and reads

$$H = \omega_L F_z - \Omega_R F_x \qquad (8.12)$$

where $\omega_L = \gamma_F B_0$ and $\Omega_R = \gamma_F B_1/2$ are the Larmor and Rabi frequencies respectively (γ_F is the Landé g-factor of the level F).

By considering the hyperfine level F=4 only and after introducing longitudinal and transverse relaxation rates γ_1 and γ_2, one obtains 9 equations for the time evolution of the populations

$$\dot{p}_M = \dot{\rho}_{M,M} = -i V_{M,M+1}(\rho_{M+1,M} - \rho_{M,M+1})$$
$$- i V_{M,M-1}(\rho_{M-1,M} - \rho_{M,M-1}) - \gamma_1(\rho_{M,M} - \rho_{M,M}^0),$$
$$(8.13)$$

where $V_{M,M'} = \langle 4, M | \Omega_R F_x | 4, M' \rangle$ and $p_M^0 = \rho_{M,M}^0$ are the steady state populations produced by the optical pumping in step 1, and 8 additional equations for the coherences

$$\dot{\rho}_{M,M-1} = -i \delta\omega \, \rho_{M,M-1}$$
$$- i V(\rho_{M-1,M-1} - \rho_{M,M}) - \gamma_2 \rho_{M,M-1}, \quad (8.14)$$

where $M = -3, ..., 4$ and $\delta\omega = \omega_L - \omega_{rf}$ is the detuning. With the complex conjugates of Eq. 8.14, the dynamics are then described by a system of 25 differential equations, which allow us to calculate numerically the steady state populations p_M and coherences $\rho_{M,M\pm 1}$. When transforming back from the rw system to the laboratory frame the $\Delta M = \pm 1$ coherences $\rho_{M,M\pm 1}$ will oscillate like $\exp[\pm i \omega_{rf} t]$.

The optical detection (step 3)

In the third step one calculates the fluorescence rate produced by absorption of the circularly polarized laser beam by the medium described by the steady

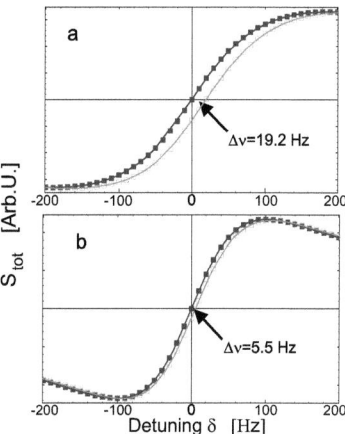

FIGURE 8.9: Calculated resonance signal produced by: (a) the single magnetic transition $|4, 4\rangle \to |4, 3\rangle$ in the case $\beta = 0$, (b) the sum of all magnetic transitions $|M\rangle \to |M - 1\rangle$ for the case $\beta = \pi/4$. Filled squares represent the $E = 0$ case, while empty squares give the magnetic resonance lineshape in an electric field of 40 kV/cm. For the numerical calculations we have assumed $\alpha_2^{(3)} = -3.49 \times 10^{-2} \; Hz/(kV/cm)^2$, which is a weighted average of previous measurements. The other parameters are $\gamma_p = 2500 \; s^{-1}$, $\gamma_1 = 1 \; s^{-1}$, $\gamma_2 = 4 \; s^{-1}$ and $\Omega_R = 50 \; s^{-1}$. Note that the different widths of (a) and (b) are consequences of different power broadenings of the single magnetic resonances.

state density matrix obtained in step 2. The time dependent signals oscillating at ω_{rf} are determined by the $\Delta M = \pm 1$ coherences $\rho_{M,M\mp 1}$. Their contribution to the absorption signal is given by

$$S_{M,M-1} \propto$$

$$\text{Re}\left[\sum_{f,m} \langle m | \mathbf{d} \cdot \mathbf{e} | M \rangle \rho_{M,M-1} \langle M - 1 | (\mathbf{d} \cdot \mathbf{e})^\dagger | m \rangle \right],$$

$$(8.15)$$

where $\mathbf{d}$ is the electric dipole operator and $\mathbf{e}$ the optical field vector. The state vectors $|M\rangle$ and $|m\rangle$ denote the states $|6S_{1/2}, F, M\rangle$ and $|6P_{1/2}, f, m\rangle$ respectively.

The effect of the tensor polarizability is taken into account by adding the differential Stark shift of the levels $|M\rangle$ and $|M-1\rangle$ to the detuning via

$$\delta = \omega_L - \omega_{rf} + \frac{3}{56}(2M-1)\alpha_2^{(3)}\mathbb{E}^2. \quad (8.16)$$

In this way one can calculate the absorptive and dispersive resonance signals in the M_x geometry by adding the contributions of all the individual transitions

$$S_{\text{tot}}(\delta) = \sum_M S_{M,M-1}(\delta). \quad (8.17)$$

The equivalent signals obtained in the M_z geometry can be calculated in an analogous way by assuming all the initial population to be concentrated in the $|4,4\rangle$ state. In Fig. 8.9 we compare the effect of an electric field of 40 kV/cm on the magnetic resonance spectra recorded in the M_z and in the M_x geometries for a particular set of the parameters, $\gamma_1 = 1\ s^{-1}$, $\gamma_2 = 4\ s^{-1}$, $\gamma_p = 2500\ s^{-1}$ [13, 15] and $\Omega_R = 50\ s^{-1}$ (corresponding to $B_1 \approx 4.5$ nT). One sees that in this case the Stark shift obtained in the M_x geometry is reduced by a factor ϵ, which, for the set of parameters $\gamma_p = 2500\ s^{-1}$ and $\Omega_R = 50\ s^{-1}$ has the value ϵ=19.2 Hz / 5.5 Hz = 3.49. We take this reduced sensitivity into account by writing the electric field dependence of the Stark shift in the M_x geometry, in analogy to Eq. 8.5, as

$$\Delta\nu_4 = \eta_x \mathbb{E}^2 = -\frac{3}{8}\frac{1}{\epsilon}\alpha_2^{(3)}(4)\mathbb{E}^2, \quad (8.18)$$

or, equivalently

$$\alpha_2^{(3)}(4) = -\frac{8}{3}\epsilon\,\eta_x. \quad (8.19)$$

In this way we obtain from (8.10)

$$\alpha_2^{(3)}(4) = (-4.36 \pm 0.28) \times 10^{-2}\frac{Hz}{(kV/cm)^2}. \quad (8.20)$$

This value is shown in Fig. 8.11 as point (e). It is in good agreement with the experimental result obtained in the M_z geometry (point d). The error bar of point (e) takes a slight uncertainty of the correction factor ϵ into account. The value of ϵ used above was obtained using our best possible estimation of the experimental parameters γ_p and Ω_R. In order to check the sensitivity of ϵ to the uncertainties of these parameters we have varied the parameters in the simulation calculation. The results shown in Fig. 8.10 indicate that ϵ is rather insensitive to parameter variations. A change of the Rabi frequency by $\pm 50\%$ changes ϵ by approximately 3%, while a change of the pumping rate γ_p by $\pm 50\%$ changes ϵ by 0.8%. We have taken this uncertainty into account by assigning a (conservative) uncertainty of 1% to ϵ, a value which does not affect the error given in Eq. 8.20.

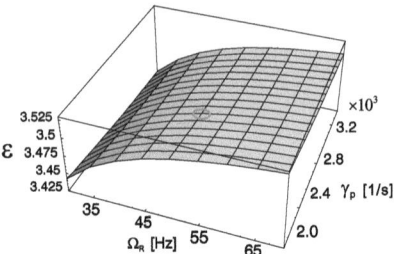

FIGURE 8.10: Correction factor ϵ as a function of the relevant parameters γ_p and Ω_R. The circle shows the parameters used to produce Fig.8.9.

8.6 Comparison with theory

Recently we have shown [8] that the inclusion of off-diagonal hyperfine matrix elements in the third order theory of forbidden tensor polarizabilities leads to a good agreement between experimental and theoretical values in the case of free Cs atoms. The present experiments show that the modulus of the tensor polarizability of Cs in solid He is approximately 10% larger than the corresponding vacuum value (Fig. 8.11). This is due to the interaction of the Cs atom with the He matrix which affects both the Cs energies and wave functions entering the third order perturbation theory. We have therefore extended our tensor polarizability calculations by including the effect of the helium matrix [9] in the frame of the so-called extended atomic bubble model [12]. The result of that calculation (details of which will be presented elsewhere [9]) is shown on the right side of Fig. 8.11 as dashed line and shows an excellent agreement with the experimental results presented above.

8.7 Summary

We have performed the first measurement of the Stark effect in the ground state of Cs atoms implanted in a solid ^{4}He matrix [19]. Measurements performed in two different experimental configurations have yielded consistent values for the forbidden tensor polarizability. The experimental results are well described by a bubble model calculation and show that the helium matrix changes the tensor polarizability by approximately 10%.

We have also measured the relative sign of the polarizabilities in the two hyperfine levels. The experimental result agrees with our theoretical pre-

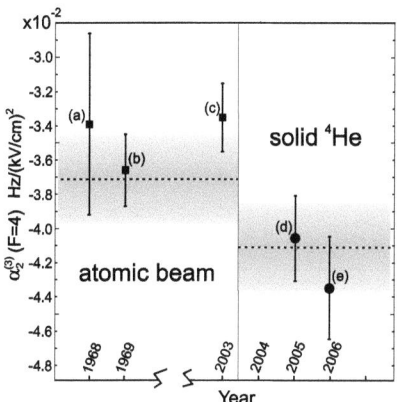

FIGURE 8.11: The Cs tensor polarizability $\alpha_2^{(3)}(F=4)$. Atomic beam measurements of (a) Carrico et al. [4], (b) Gould et al. [5], (c) Ospelkaus et al. [6]. Points (d) and (e) represent the measurements in solid helium, in the M_z and M_x geometries respectively, reported in this work. The dashed lines are the theoretical values for the free atom [8] and for Cs in a solid helium matrix [9], together with their uncertainties (shaded bands).

diction and is in contradiction with the sign predicted by a previous calculation. This confirms the need [8] for a reevaluation of the dynamic Stark shift of primary frequency standards induced by the black-body radiation field [20], when this effect is inferred from static Stark shift measurements as done, e.g., by Simon et al. [21].

ACKNOWLEDGMENTS We acknowledge the skillful help of Stephan Gröger for the measurements in the M_x geometry. This work was supported by the grant number 200020-103864 of the Swiss National Science Foundation.

References

[1] J. R. P. Angel and P. G. H.Sandars, Proc. R. Soc. London, A **305**, 125 (1968).

[2] R. D. Haun and J. R. Zacharias, Phys. Rev. **107**, 107 (1957).

[3] E. Lipworth and P. G. H. Sandars, Phys. Rev. Lett. **13**, 716 (1964).

[4] J. P. Carrico, A. Adler, M. R. Baker, S. Legowski, E. Lipworth, P. G. H. Sandars, T. S. Stein, and C. Weisskopf, Phys. Rev. **170**, 64 (1968).

[5] H. Gould, E. Lipworth, and M. C. Weisskopf, Phys. Rev. **188**, 24 (1969).

[6] C. Ospelkaus, U. Rasbach, and A. Weis, Phys. Rev. A **67**, 011402 (2003).

[7] P. G. H. Sandars, Proc. Phys. Soc. **92**, 857 (1967).

[8] S. Ulzega, A. Hofer, P. Moroshkin, and A. Weis, Europhys. Lett. **76** (2006).

[9] A. Hofer, P. Moroshkin, S. Ulzega, and A. Weis, in preparation.

[10] M. Arndt, S. I. Kanorsky, A. Weis, and T. W. Hänsch, Phys. Lett. A **174**, 298 (1993).

[11] A. Weis, S. Kanorsky, S. Lang, and T. W. Hänsch, in *Atomic Physics Methods in Modern Research* (Springer, 1997), vol. 499 of *Lecture Notes in Physics*, pp. 57–75.

[12] P. Moroshkin, A. Hofer, S. Ulzega, and A. Weis, Low Temp. Phys. **32**, 1297 (2006).

[13] M. Arndt, S. Kanorsky, A. Weis, and T. W. Hänsch, Phys. Rev. Lett. **74**, 1359 (1995).

[14] S. Kanorski, S. Lang, S. Lücke, S. B. Ross, T. W. Hänsch, and A. Weis, Phys. Rev. A **54**, 1010 (1996).

[15] S. Lang, S. Kanorsky, T. Eichler, R. Müller-Siebert, T. W. Hänsch, and A. Weis, Phys. Rev. A **60**, 3867 (1999).

[16] R. Müller-Siebert, Ph.D. thesis, University of Fribourg, Switzerland (2003), unpublished.

[17] S. Gröger, G. Bison, J. L. Schenker, R. Wynands, and A. Weis, Eur. Phys. J. D **38**, 239 (2006).

[18] D. Budker, W. Gawlik, D. F. Kimball, S. M. Rochester, V. V. Yashchuk, and A. Weis, Rev. Mod. Phys. **74**, 1153 (2002).

[19] A preliminary analysis of the data discussed here was shown in [8]. That analysis did not take the thermal expansion of the field plate spacing into account.

[20] W. M. Itano, L. L. Lewis, and D. J. Wineland, Phys. Rev. A **25** (1982).

[21] E. Simon, P. Laurent, and A. Clairon, Phys. Rev. A **57**, 436 (1998).

Chapter 9

Paper VI:
Calculation of the forbidden electric tensor polarizabilities of free Cs atoms and of Cs atoms trapped in a solid ^{4}He matrix

> This paper gives more details of the calculation in third order perturbation used for the evaluation of the tensor polarizability in the Cs ground state. Parts of the theory were published elsewhere (Ulzega et al., Europhys. Lett. 76, 1074 (2006)). The slightly different experimental value for the tensor polarizability for Cs implanted in solid ^{4}He reported here coincides well with an extended model that takes the influence of the He matrix into account.
>
> My main contributions to the work were:
>
> - Setting up computer code for solving the Schrödinger equation with a scaled Thomas-Fermi model potential to calculate the wavefunctions and energy levels of the free Cs atom. Analyzing the model parameters to fit experimental energies. Use of the wavefunctions to evaluate transition matrix elements (dipole and hyperfine) for the numerical calculation of the third order perturbation expansion. Estimation of the effect of the continuum states by explicit calculation of continuum state wavefunctions.
>
> - Development and use of the extended bubble model to calculate the influence of the He matrix on the tensor polarizability.
>
> - Producing figures, graphs and text for the paper.

Calculation of the forbidden electric tensor polarizabilities of free Cs atoms and of Cs atoms trapped in a solid ^{4}He matrix

A. Hofer[1], P. Moroshkin[1], S. Ulzega[2] and A. Weis[1]

[1]*Département de Physique, Université de Fribourg, Chemin du Musée 3, 1700 Fribourg, Switzerland*
[2]*EPFL, Lausanne, Switzerland.*

submitted to Phys. Rev. A

Abstract: We give a detailed account on our semi-empirical calculations of the forbidden electric tensor polarizability $\alpha_2^{(3)}$ of the ground state of free Cs atoms and of Cs atoms implanted in a solid ^{4}He matrix. The results are compared with measurements of $\alpha_2^{(3)}$ in the free atoms [1] and in He-trapped atoms [2]. Novel features with respect to calculations by other authors are the inclusion of off-diagonal hyperfine interactions and an analysis of contributions from continuum states, which turn out to be negligible. For both samples the results of the calculations are in good agreement with the experimental values, thereby settling a long-standing discrepancy.

9.1 Introduction

The interaction of an atom with an external static electric field (Stark effect) is one of the fundamental interactions in atomic physics. In atoms with degenerate orbital momentum states of opposite parity the change of level energies induced by the electric field is linear in the field strength $\mathbb{E}$, the hydrogen atom being the most prominent example. In most atoms, however, the Stark shift is quadratic in the field strength, and the electric field induced shift of a magnetic hyperfine (hf) sublevel $|\gamma\rangle = |n L_J, F, M\rangle$ is commonly parameterized in terms of the (electrostatic) polarizability $\alpha(\gamma)$ as

$$\Delta E(\gamma) = -\frac{1}{2}\alpha(\gamma)\mathbb{E}^2. \qquad (9.1)$$

In second order perturbation theory, the polarizability of states nL_J with $J \geq 1$ can be written as the sum of an F- and M-independent scalar polarizablity $\alpha_0^{(2)}$ and an F- and M-dependent tensor polarizability $\alpha_2^{(2)}$, while states with $J < 1$ have only a scalar polarizability. For spherically symmetric states, such as the $nS_{1/2}$ and $nP_{1/2}$ states in alkali atoms, the Stark effect is thus a purely scalar effect, meaning that all hyperfine sublevels $|n L_{1/2}, F, M\rangle$ experience the same common shift. However, an F- and M-dependence of the Stark shifts in the $|nS_{1/2}\rangle$ alkali ground states was already found experimentally in the late 1950's and early 1960's [3, 4], and thus indicated the existence of a (forbidden) tensor polarizability of those states.

In 1967 Sandars [5] could show that this forbidden polarizability can be explained by expanding the perturbation treatment to third order, considering the Stark and the hyperfine interactions as simultaneous perturbations. The corresponding third order polarizability can also be expressed in terms of an F-dependent third order scalar polarizability $\alpha_0^{(3)}(F)$ and an F- and M-dependent third order tensor polarizability $\alpha_2^{(3)}(F, M)$. The polarizability in Eq. (9.1) can thus be written as

$$\begin{aligned}\alpha(\gamma) &= \alpha_0^{(2)}(nL_J) + \alpha_2^{(2)}(nL_J, F, M) \\ &+ \alpha_0^{(3)}(nL_J, F) + \alpha_2^{(3)}(nL_J, F, M)\end{aligned}(9.2)$$

where the superscripts refer to the order of perturbation, while the subscripts refer to the rotational symmetry (scalar or second rank tensor) of the interaction.

Sandars' expressions were evaluated numerically in [4, 6] under simplifying assumptions, and yielded values of $\alpha_2^{(3)}(F, M)$ for five alkali isotopes whose (absolute) calculated values were systematically larger than the corresponding experimental values [7, 6, 8, 1]. Recently we have remeasured the third order tensor polarizability $\alpha_2^{(3)}(F, M)$ of the Cs ground state in an all-optical atomic beam experiment [8], yielding good agreement with the measurements from the 1960's [6, 7]. We have also measured $\alpha_2^{(3)}(F, M)$ of cesium atoms implanted in the cubic phase of a ^{4}He crystal, yielding a value whose modulus is $\approx$10% larger than in the free atom [2]. In parallel we have reanalyzed and extended the theoretical expressions of the tensor polarizabilities [1]. We have further extended the third order perturbation calculation [5] by including off-diagonal hyperfine interactions, not considered in the earlier calculations, as well as contributions from higher lying bound and continuum states. This has yielded a theoretical value for $\alpha_2^{(3)}$ of Cs($6S_{1/2}$) which is in good agreement with all existing experimental results.

As discussed in [1] we have uncovered in an earlier calculation [5] an error concerning the relative signs of the polarizabilities of the two ground state hyperfine levels. Recently we have given an experimental verification of the sign of our calculation and discussed its implication for the dynamic Stark shift of primary frequency standards by the black-body radiation field [2].

In this paper we present more details of that calculation, whose results were already outlined in [1] and [2]. In addition we present a calculation of $\alpha_2^{(3)}$ for cesium atoms in a cubic ^{4}He crystal by evaluating the matrix-induced alterations of the atomic energies and of the electric dipole and hyperfine matrix elements in the frame of the so-called standard bubble model [9]. Here too we obtain an excellent agreement with the experimental values.

9.2 Theory

The Stark effect, i.e., the interaction of an alkali atom with an external electric field $\mathbb{E}$ is described by the Hamiltonian $H_{St} = -\vec{d}\cdot\vec{\mathbb{E}}$, where $\vec{d}$ is the electric dipole operator of the valence electron. Because of parity conservation, the Stark interaction vanishes in first order, and the effect of the electric field on the atomic level structure appears only in the next higher order(s), yielding shifts that are quadratic in the applied field strength.

9.2.1 Second order perturbation theory

The second order energy perturbation of the magnetic hyperfine sublevel $|\beta\rangle = |6S_{1/2}, F, M\rangle$ of the ground state is given by

$$\Delta E^{(2)}(\beta) = \sum_\gamma \frac{|\langle\beta|\,\mathrm{H}_{\mathrm{St}}\,|\gamma\rangle|^2}{E_\beta - E_\gamma}, \qquad (9.3)$$

where, according to the selection rule $\Delta L = \pm 1$ imposed by the Stark operator, the sum is to be taken over all excited P-states $|\gamma\rangle = |nP_J, f, m\rangle$, including continuum states, and where E_γ are the unperturbed energies of those states. Following [10] one can define an effective second order Stark operator

$$_2\mathrm{H}_{\mathrm{eff}} = \left(\vec{d}\cdot\vec{\mathrm{E}}\right)\,_2\lambda\left(\vec{d}\cdot\vec{\mathrm{E}}\right), \qquad (9.4)$$

in which the scalar projection operator $_2\lambda$ is defined as

$$_2\lambda = \sum_\gamma \frac{|\gamma\rangle\langle\gamma|}{E_\beta - E_\gamma}. \qquad (9.5)$$

With this definition of $_2\mathrm{H}_{\mathrm{eff}}$ the second order energy shift $\Delta E^{(2)}$ of the state $|\beta\rangle$ is given by the expectation value

$$\Delta E^{(2)}(\beta) = \langle\beta|_2\mathrm{H}_{\mathrm{eff}}|\beta\rangle, \qquad (9.6)$$

similar to the expression from first order perturbation theory. Note that the superscripts on the right of $\Delta E^{(n)}$ and $\alpha^{(n)}$ and the subscripts on the left of $_n\mathrm{H}_{\mathrm{eff}}$ and $_n\lambda$ refer to the order of perturbation.

The interaction Hamiltonian can be factorized [10] into an electric field-dependent part and a part which depends only on atomic properties by defining the components of two rank-K tensor operators as

$$\begin{aligned}
\mathcal{E}_Q^{(K)} &= [\mathrm{E}\otimes\mathrm{E}]_Q^{(K)} \\
&= \sum_{q,q'}\langle 11qq'|KQ\rangle\,\mathrm{E}_q^{(1)}\mathrm{E}_{q'}^{(1)} \qquad (9.7) \\
\mathcal{D}_Q^{(K)} &= [d^{(1)}\otimes\,_2\lambda\otimes d^{(1)}]_Q^{(K)} \\
&= \sum_{q,q'}\langle 11qq'|KQ\rangle\,d_q^{(1)}\,_2\lambda\,d_{q'}^{(1)}, \qquad (9.8)
\end{aligned}$$

where the $\langle 11qq'|KQ\rangle$ are Clebsch-Gordan coefficients and where $d_q^{(1)}$ and $\mathrm{E}_q^{(1)}$ are the spherical components of the dipole operator and the electric field, respectively. The effective Stark operator can then be written as the sum

$$_2\mathrm{H}_{\mathrm{eff}} = \sum_{K=0}^{2}(-1)^K\,_2\mathrm{H}_{\mathrm{eff}}^{(K)} = \,_2\mathrm{H}_{\mathrm{eff}}^{(0)} + \,_2\mathrm{H}_{\mathrm{eff}}^{(2)} \qquad (9.9)$$

of its multipole components

$$_2\mathrm{H}_{\mathrm{eff}}^{(K)} = \sum_Q (-1)^Q \mathcal{E}_Q^{(K)}\mathcal{D}_{-Q}^{(K)}. \qquad (9.10)$$

The vector contribution vanishes since $\mathcal{E}^{(1)} \propto [\mathrm{E}\otimes\mathrm{E}]^{(1)} \propto \vec{E}\times\vec{E}$. The scalar term $\mathcal{E}^{(0)} \propto [\mathrm{E}\otimes\mathrm{E}]^{(0)} \propto \vec{\mathrm{E}}\cdot\vec{\mathrm{E}} = \mathrm{E}^2$ depends only on the magnitude of the field, while the rank-2 tensor term $_2\mathrm{H}_{\mathrm{eff}}^{(2)}$ depends on its orientation, as can be seen, e.g., from its $Q=0$ component $\mathcal{E}_0^{(2)} \propto 3\mathrm{E}_z^2 - \mathrm{E}^2 = \mathrm{E}^2(3\cos^2\theta - 1)$. With this notation, the second order Stark effect $\Delta E^{(2)}(\beta)$ can be written as the expectation value

$$\Delta E^{(2)}(\beta) = \langle\beta|\,_2\mathrm{H}_{\mathrm{eff}}^{(0)} + \,_2\mathrm{H}_{\mathrm{eff}}^{(2)}|\beta\rangle. \qquad (9.11)$$

The Wigner-Eckart theorem implies that the matrix element $\langle nLJ|_2\mathrm{H}_{\mathrm{eff}}^{(K)}|nLJ\rangle$ vanishes unless $0 \leq K \leq 2J$, so that the tensor part $(K=2)$ of the interaction vanishes

$$\langle\beta|\,_2\mathrm{H}_{\mathrm{eff}}^{(2)}|\beta\rangle \propto \langle 6S_{1/2}\,\|\,_2\mathrm{H}_{\mathrm{eff}}^{(2)}\,\|\,6S_{1/2}\rangle \equiv 0 \quad (9.12)$$

for the spherically symmetric $6S_{1/2}$ state. As a consequence, the Stark effect in the alkali ground state treated in second order perturbation theory is a purely scalar effect which is parameterized in terms of the F- and M-independent second order scalar polarizability $\alpha_0^{(2)}$ as

$$\Delta E^{(2)}(\beta) = -\frac{1}{2}\alpha_0^{(2)}\mathrm{E}^2. \qquad (9.13)$$

This energy perturbation results in an overall shift of the ground state sublevels. The experimental value of $\alpha_0^{(2)}(6S_{1/2})$ is $9.98(2) \times 10^3$ Hz/(kV/cm)^2 [11], and is theoretically well understood at a level of 10^{-3}[12, 13]. The F- and M-dependent shifts of the ground state sublevels discussed below are approximately 5 and 7 orders of magnitude smaller than this global scalar shift.

9.2.2 Third order perturbation theory

As already mentioned, Sandars has shown [5] that the forbidden electric field induced lifting of the Zeeman degeneracy can be explained by extending the perturbation theory to third order, including simultaneously the Stark interaction and the hyperfine interactions in terms of the perturbation operator $W = \mathrm{H}_{\mathrm{St}} + H_{\mathrm{hf}}^{\mathrm{Fc}} + H_{\mathrm{hf}}^{\mathrm{dd}} + H_{\mathrm{hf}}^{\mathrm{q}}$, which consists of the Stark (H_{St}), the hyperfine Fermi contact ($H_{\mathrm{hf}}^{\mathrm{Fc}}$), the hyperfine dipole-dipole ($H_{\mathrm{hf}}^{\mathrm{dd}}$) and the hyperfine quadrupole ($H_{\mathrm{hf}}^{\mathrm{q}}$) operators. In [1] we have given explicit expressions of the hyperfine operators in terms of irreducible tensor operators.

Following the general rules of perturbation theory the third order energy perturbation of the ground state level $|\beta\rangle$ is given by

$$\Delta E^{(3)}(\beta) = \sum_{\gamma\neq\beta,\delta\neq\beta} \frac{\langle\beta|W|\gamma\rangle\langle\gamma|W|\delta\rangle\langle\delta|W|\beta\rangle}{(E_\beta - E_\gamma)(E_\beta - E_\delta)}$$
$$- \langle\beta|W|\beta\rangle \sum_{\gamma\neq\beta} \frac{|\langle\gamma|W|\beta\rangle|^2}{(E_\beta - E_\gamma)^2}, \quad (9.14)$$

where E_γ and E_δ are the unperturbed state energies.

The two terms of Eq. (9.14) are trilinear forms of the perturbation operator W, of which only terms proportional to $\mathbb{E}^2$ give nonzero contributions to the Stark interaction. As discussed by Ulzega et al. [1] the ground state hyperfine interaction $\mathrm{E_{hf}}(6S)$ is factored out from the sum over pure Stark matrix elements in the second term of Eq. (9.14). This leads to M-independent, but F-dependent shifts of the ground state levels which can be parametrized, according to Eqs. (9.1) and (9.2), by a third order scalar polarizability $\alpha_0^{(3)}(F)$. The second term thus does not contribute to the tensor polarizability, but gives the leading contribution (type A interactions in the notation of [1]) to the Stark shift of the hyperfine (clock transition) frequency [14, 15]. The contribution from this term is thus suppressed by a factor on the order of $\mathrm{E_{hf}}(6S)/\Delta E_{6P-6S} \approx 10^{-5}$ with respect to the second order scalar polarizability $\alpha_0^{(2)}$.

As discussed by Ulzega et al. [1] the first term of Eq. (9.14) has contributions from both diagonal and off-diagonal hyperfine matrix elements. The diagonal contributions (type B interactions in the notation of [1]) dominate (see Fig. 9.1) and are suppressed by a factor on the order of $\mathrm{E_{hf}}(6P)/\Delta E_{6P-6S} \approx 10^{-7}$ with respect to the second order scalar shift $\alpha_0^{(2)}$. The Zeeman splitting of the ground state levels by the electric field is thus approximately 100 times smaller than the shift of the clock transition frequency.

The third order perturbation can again be expressed in terms of an effective Hamiltonian

$$_3\mathrm{H}_\mathrm{eff} = \left(\vec{d}\cdot\vec{\mathbb{E}}\right) {_3\lambda} \left(\vec{d}\cdot\vec{\mathbb{E}}\right), \quad (9.15)$$

in which the projection operator $_3\lambda$ is given by

$$_3\lambda = \sum_\gamma \frac{|\gamma\rangle\langle\gamma|\,H_\mathrm{hf}\,|\gamma\rangle\langle\gamma|}{(E_\beta - E_\gamma)^2}. \quad (9.16)$$

The effective Hamiltonian can be expressed as the sum of a scalar and a rank-2 tensor, yielding the third order energy perturbation

$$\Delta E^{(3)}(\beta) = \langle\beta|\,_3\mathrm{H}_\mathrm{eff}^{(0)} + {_3\mathrm{H}_\mathrm{eff}^{(2)}}\,|\beta\rangle, \quad (9.17)$$

in a form equivalent to Eq. (9.11). The scalar part turns out to have the same F-dependence as the third order scalar polarizability, $\alpha_0^{(3)}(F)$, from the second term of (9.14) and gives a correction to the latter on the order of 1%.

By applying the Wigner-Eckart theorem, the contribution of the second rank tensor part can be written as

$$\langle\beta|\,_3\mathrm{H}_\mathrm{eff}^{(2)}\,|\beta\rangle \propto \left[3M^2 - F(F+1)\right](3\mathbb{E}_z^2 - \mathbb{E}^2)$$
$$\langle F \parallel [d^{(1)}\otimes {_3\lambda}\otimes d^{(1)}]^{(2)} \parallel F\rangle. \quad (9.18)$$

The electronic and nuclear angular momenta in this equation cannot be decoupled since the operator $[d^{(1)}\otimes {_3\lambda}\otimes d^{(1)}]^{(2)}$ depends explicitly on the hyperfine interaction term $\mathbf{J}\cdot\mathbf{I}$ and one can state in general that $\langle F \parallel [d^{(1)}\otimes {_3\lambda}\otimes d^{(1)}]^{(2)} \parallel F\rangle \neq 0$ for states with $F \geq 1$. The tensor part has therefore an explicit F- and M-dependence which can be parameterized in terms of a third order tensor polarizability $\alpha_2^{(3)}(6S_{1/2}, F, M)$.

We can summarize the above results by parameterizing the two terms of the third order interaction [Eq. (9.14)] by a total third order polarizability as

$$\Delta E^{(3)} = -\frac{1}{2}\alpha^{(3)}(F,M)\mathbb{E}^2, \quad (9.19)$$

with

$$\alpha^{(3)}(F,M) = \alpha_0^{(3)}(F) + \alpha_2^{(3)}(F)\frac{3M^2 - F(F+1)}{2I(2I+1)}f(\theta). \quad (9.20)$$

The function $f(\theta) = 3\cos^2\theta - 1$ expresses the dependence of $\alpha^{(3)}$ on the angle θ between the electric field and the quantization axis.

9.3 The third order polarizability of the free cesium atom

9.3.1 Earlier calculation revisited

The leading term in the perturbation sum (9.14) is given by

$$\Delta E^{(3)}(6S_{1/2},F,M) = \sum_{n,J,F,m} \langle nP_J,F|\,H_\mathrm{hf}\,|nP_J,F\rangle$$
$$\times \frac{|\langle nP_J,F,m|\,\mathrm{H_{St}}\,|6S_{1/2},F,M\rangle|^2}{(E_{nP_J,F,m} - E_{6S_{1/2},F,M})}, \quad (9.21)$$

9.3 The third order polarizability of the free cesium atom

where we have used the fact that H_{hf} does not mix F values, and where we have allowed for M-mixing by H_{St} in case the electric field is not along the quantization axis. This term is diagrammatically represented as diagram B in [1], and we shall refer to it below as type B interaction. The corresponding tensor polarizability of the $F=4$ state of Cs was evaluated in [6] considering – among other simplifying assumptions discussed in [1] – only diagonal hyperfine matrix elements in the $6P_{1/2}$ and $6P_{3/2}$ states. The result of that calculation is shown as point (f) in Fig. 9.4 and is in disagreement with the experimental results. We have redone this calculation after dropping all simplifying assumptions of [6], while still considering only diagonal matrix elements. We also used recent more precise experimental values [16] for the reduced dipole matrix elements $\langle 6S_{1/2} \| d \| 6P_J \rangle$ which yield the leading contribution. Point (g) in Fig. 9.4 represents the result of this reanalysis. As a result, the gap between theoretical and experimental values of the Cs tensor polarizability increases. Extending the perturbation sum to nP_J states with $n>6$ does not affect the discrepancy at a significant level (line B in Table 9.5 and Fig. 9.1).

9.3.2 Inclusion of off-diagonal hf matrix elements

All previous third order calculations have considered only diagonal hyperfine matrix elements (diagram **B** in Fig. 1 of [1]). However, the first term of the general third order expression [Eq. (9.14)] contains also off-diagonal hyperfine matrix elements. In [1] we have identified five different types of off-diagonal hyperfine mixing contributions:

- Type 1: n-mixing of $nS_{1/2}$ states.
- Type 2: n-mixing of nP_J states with given J.
- Type 3: J-mixing of nP_J states with given n.
- Type 4: n and J mixing of nP_J states.
- Type 5: mixing of $mD_{3/2}$ and $6S_{1/2}$ states.

Note that the type 1 interactions have an F, but no M dependence and contribute thus only to the clock frequency shift. The different mixing types are represented as diagrams in Fig. 1 of [1] and the labelling used here follows the notation of that figure. The energy shifts due to interactions of type 2, 3, and 4 are given by sums of terms with the general structure

$$\frac{\langle \beta | H_{\text{St}} | \gamma_1 \rangle \langle \gamma_1 | H_{\text{hf}} | \gamma_2 \rangle \langle \gamma_2 | H_{\text{St}} | \beta \rangle}{(E_\beta - E_{\gamma_1})(E_\beta - E_{\gamma_2})}, \quad (9.22)$$

with $|\beta\rangle = |nS_{1/2}, F, M\rangle$, $|\gamma_2\rangle = |nP_J, f, m\rangle$, and $|\gamma_1\rangle = |n'P_{J'}, f, m\rangle$. The contributions to the interaction of type 5 have the form

$$\frac{\langle \beta | H_{\text{St}} | \delta \rangle \langle \delta | H_{\text{St}} | \gamma \rangle \langle \gamma | H_{\text{hf}} | \beta \rangle}{(E_\beta - E_\delta)(E_\beta - E_\gamma)}, \quad (9.23)$$

where $|\gamma\rangle = |mD_{3/2}, F, M\rangle$ and $|\delta\rangle = |nP_J, f, m\rangle$.

Some of the off-diagonal hyperfine matrix elements were considered earlier in calculations [14, 15, 17] of the Stark shift of the $|6S_{1/2}, F=3, M=0\rangle \rightarrow |6S_{1/2}, F=4, M=0\rangle$ clock transition, but were never considered in a calculation of the tensor polarizability. The off-diagonal (type 1) matrix elements of the Fermi contact interaction between $S_{1/2}$ states contribute only to the clock frequency shift and are of no relevance here.

The (diagonal) matrix elements of the hyperfine quadrupole interaction H_{hf}^{q} are only relevant for states with $L, J > 1/2$ and their numerical values are two orders of magnitude smaller than all other hyperfine matrix elements. The matrix elements of the quadrupole part of the hyperfine Hamiltonian can be decomposed in a similar way. However, since their numerical evaluation yields only a relative contribution of 10^{-3} to the tensor polarizability we do not reproduce the corresponding algebraic expressions here. The only hyperfine operator contributing to the off-diagonal matrix elements is the dipole-dipole operator $H_{\text{hf}}^{\text{dd}}$, which can be expressed in terms of irreducible tensor operators as

$$H_{\text{hf}}^{\text{dd}} = a(r) \left(\mathbf{L}^{(1)} - \sqrt{10} [\mathbf{C}^{(2)} \otimes \mathbf{S}^{(1)}]^{(1)} \right) \cdot \mathbf{I}^{(1)}, \quad (9.24)$$

where $\mathbf{L}^{(1)}$, $\mathbf{S}^{(1)}$, and $\mathbf{I}^{(1)}$ are the irreducible vector operators associated with the orbital angular momentum, the electronic spin, and the nuclear spin, respectively, and $\mathbf{C}^{(k)} = \sqrt{\frac{4\pi}{2k+1}} \mathbf{Y}^{(k)}$ are the normalized spherical harmonic operators of rank k. The radial dependence of the hyperfine operator is given by $a(r)$. The two terms in (9.24) correspond to the dipolar magnetic interaction of the nuclear spin $\mathbf{I}$ with the orbital ($\propto \mathbf{L}^{(1)}$) and the electronic spin ($\propto \mathbf{C}^{(2)} \times \mathbf{S}^{(1)}$), respectively. The two interactions obey the selection rules $\Delta L = 0$ and $\Delta L = 0, \pm 2$, respectively. The relevant $\Delta L = 0$ interactions (types 2,3,4) have non-vanishing off-diagonal matrix elements of the form $\langle \gamma_1 | H_{\text{hf}}^{\text{dd}} | \gamma_2 \rangle$ where $|\gamma_i\rangle = |n_i P_{J_i}, f, m\rangle$. The explicit reduction of those matrix elements leads to

$\langle \gamma_1 | H_{\text{hf}}^{\text{dd}} | \gamma_2 \rangle =$
$(-1)^{f+L_1+J_2+1} 3\sqrt{7}\sqrt{(2J_1+1)(2J_2+1)}$
$\begin{Bmatrix} J_1 & 7/2 & f \\ 7/2 & J_2 & 1 \end{Bmatrix} (\epsilon^{\text{orbital}} + \epsilon^{\text{spin}}) \langle a(r) \rangle ,$

(9.25)

where we have separated the contribution of the orbital magnetic dipole interaction

$\epsilon^{\text{orbital}} = (-1)^{J_2} 2\sqrt{3} \begin{Bmatrix} L_1 & J_1 & 1/2 \\ J_2 & L_2 & 1 \end{Bmatrix} \delta_{L_1,L_2} ,$

(9.26)

from that of the spin dipolar interaction

$\epsilon^{\text{spin}} = (-1)^{7/2} 3\sqrt{10}\sqrt{(2L_1+1)(2L_2+1)}$
$\begin{pmatrix} L_1 & 2 & L_2 \\ 0 & 0 & 0 \end{pmatrix} \begin{Bmatrix} L_1 & L_2 & 2 \\ 1/2 & 1/2 & 1 \\ J_1 & J_2 & 1 \end{Bmatrix} .$ (9.27)

The coupling constant is given by

$\langle a(r) \rangle = \int_0^\infty \Psi_{n_1 P_{J_1}}^* a(r) \Psi_{n_2 P_{J_2}} r^2 dr$
$= \frac{2g_I}{h} \frac{\mu_0}{4\pi} \mu_b^2 \int_0^\infty \Psi_{n_1 P_{J_1}}^* \frac{1}{r^3} \Psi_{n_2 P_{J_2}} r^2 dr .$

(9.28)

Because of the second rank tensor character of the spherical harmonic operator $\mathbf{C}^{(2)}$ in Eq. (9.24), the spin dipolar term can also couple states with $\Delta L = \pm 2$ and can thus contribute to the third order Stark effect with off-diagonal matrix elements of the form $\langle \gamma | H_{\text{hf}}^{\text{dd}} | \beta \rangle$, given by Eq. 9.23, where $|\gamma\rangle = |mD_{3/2}, F, M\rangle$ with $m \geq 5$ and $|\beta\rangle = |6S_{1/2}, F, M\rangle$. In the latter case $\epsilon^{\text{orbital}}$ vanishes and Eq. 9.25 reduces to

$\langle \gamma | H_{\text{hf}}^{\text{dd}} | \beta \rangle =$
$\frac{\sqrt{5-F}\sqrt{F-2}\sqrt{F+3}\sqrt{F+6}}{4} \langle a(r) \rangle ,$ (9.29)

where the corresponding coupling constant is given by

$\langle a(r) \rangle = \frac{2g_I}{h} \frac{\mu_0}{4\pi} \mu_b^2 \int_0^\infty \Psi_{mD_{3/2}}^* \frac{1}{r^3} \Psi_{6S_{1/2}} r^2 dr .$

(9.30)

9.3.3 Electric dipole matrix elements

Besides the hyperfine matrix elements the expressions involve matrix elements of the Stark interaction H_{St}, i.e., of the electric dipole operator d between S and P states. The latter can be reduced by applying the Wigner-Eckart theorem and the standard angular momentum decoupling rules, leading to the reduced matrix elements

$\langle nS_{1/2} \| d \| mP_J \rangle = (-1)^{J-\frac{1}{2}} \langle mP_J \| d \| nS_{1/2} \rangle$
$= -\sqrt{\frac{2J+1}{3}} D_{\text{nS,m}P_J} ,$ (9.31)

in which the radial integral is given by

$D_{\text{nS,m}P_J} = e \int_0^\infty R_{mP_J}(r) r^3 R_{nS_{1/2}}(r) dr$ (9.32)

where R_{nL_J} are the radial wavefunctions.

We note that the phases (signs) of the reduced matrix elements [Eq. (9.31)] are irrelevant for evaluating contributions involving diagonal hyperfine interactions, but are of fundamental importance for the contributions involving off-diagonal matrix elements.

Although dipole matrix elements between low lying states in the Cs atoms can be calculated quite accurately using relativistic Hartree-Fock calculations (see, e.g., [19, 21]) we have decided to rather use (more precise) experimental values, whenever they were available. Table 9.1 lists (in bold) the reduced dipole matrix elements used in the present calculation. For the matrix elements involving higher lying states neither experimental nor theoretical values were available. In that case we have evaluated the corresponding radial integrals using non-relativistic wavefunctions obtained by solving the Schrödinger equation as described in the next paragraph. The values of Table 9.1 show that this approach reproduces the matrix elements to better than 10%. Assuming that the quoted accuracy also holds for excited states, and considering that the excited states give only a minor contribution to the final result (cf. Table 9.1), we are confident that there is no significant uncertainty introduced by using our theoretical values for excited state matrix elements.

9.3.4 Wavefunctions of the free cesium atom

The non-relativistic wavefunctions of the states $|nL, J\rangle$ can be separated into radial $R_{nL,J}(r)$ and

9.3 The third order polarizability of the free cesium atom

$nS_{1/2}, mP_J$	Experiments	Theory (this work)	Deviation
$6S_{1/2}, 6P_{1/2}$	± **5.5087(75)**[a]	-5.3982	2%
$6S_{1/2}, 6P_{3/2}$	± **5.4829(62)**[a]	-5.3009	3%
$6S_{1/2}, 7P_{1/2}$	± **0.3377(24)**[b]	-0.3092	8%
$6S_{1/2}, 7P_{3/2}$	± **0.5071(43)**[b]	-0.4617	9%
$7S_{1/2}, 6P_{1/2}$	± **5.184(27)**[c]	+5.245	1%
$7S_{1/2}, 6P_{3/2}$	± **5.611(27)**[c]	+5.696	2%
$7S_{1/2}, 7P_{1/2}$	± **12.625**[d]	-13.298	5%
$7S_{1/2}, 7P_{3/2}$	± **12.401**[d]	-13.015	5%
$6S_{1/2}, 8P_{1/2}$	–	**-0.078**	–
$6S_{1/2}, 8P_{3/2}$	–	**-0.130**	–
$6S_{1/2}, 9P_{1/2}$	–	**-0.045**	–
$6S_{1/2}, 9P_{3/2}$	–	**-0.078**	–
$6S_{1/2}, 10P_{1/2}$	–	**-0.030**	–
$6S_{1/2}, 10P_{3/2}$	–	**-0.054**	–

TABLE 9.1: Radial integrals $D_{nS_{1/2},mP_J}$ in atomic units. Experimental values from (a) Rafac et al. [16], (b) Vasilyev et al. [18], (c) experimental value quoted in [19], and (d) Bennett et al. [20], with signs determined from Eq. (9.31). The values marked in bold were used in the calculation of the third order tensor polarizability. The theoretical values calculated from Schrödinger wavefunctions were used for matrix elements involving states with $n > 7$. The theoretical values of the states involving $n = 6, 7$ were not used in the calculation but are shown merely to illustrate the accuracy of our calculation.

angular $Y_{L,m}(\theta,\phi)$ parts

$$\Psi_{nLJ}(\mathbf{r}) = \langle \mathbf{r}|nL,J\rangle = R_{nL,J}(r)\,Y_{L,m}(\theta,\phi).$$
(9.33)

The radial wavefunctions are found as solutions of the radial Schrödinger equation for $u(r) = rR_{n,J}(r)$

$$-\frac{1}{2}\frac{d^2u(r)}{dr} + \left[V_{\text{Cs}}(r) + \frac{L(L+1)}{2r^2}\right]u(r) = E\,u(r)$$
(9.34)

For the potential $V_{\text{Cs}}(r)$ we have used a scaled Thomas-Fermi model potential $V_{\text{TF}}(r,\lambda)$ (with a scaling parameter λ) including dipolar and quadrupolar core-polarization corrections $V_{\text{pol}}(r)$ and the spin-orbit interaction $V_{\text{so}}(r)$ that depends on the angular momentum L

$$V_{\text{Cs}}(r) = V_{\text{TF}}(r,\lambda) + V_{\text{pol}}(r) + V_{\text{so}}(r). \quad (9.35)$$

This model follows the work of Gombas [22] and Norcross [23] and was explained in detail in [9]. Using this approach we have calculated the electronic wavefunctions for nS_J, nP_J, and nD_J states up to $n = 200$.

9.3.5 Terms with diagonal hf matrix elements

As stated above terms involving diagonal hyperfine matrix elements (hyperfine coupling constants) in the nP_J states give the leading contribution to the third order tensor polarizability. In the numerical evaluation of the terms we use again experimental values, when they are available. In Table 9.2 we list the hyperfine coupling constants used (in bold). A comparison of experimental coupling constants of low-lying states with constants calculated using our Schrödinger wavefunctions is also made in order to illustrate the accuracy of the theoretical approach. As for the radial integrals there are no experimental data for the high lying states and we used our theoretical values, and, again, our accuracy is sufficient since it affects only the (very small) contributions from excited states.

It can be easily seen that the off-diagonal contributions are sensitive to the sign of the radial matrix elements. Whenever we extracted such matrix elements from experimental data – which all determined squared matrix elements – we have used the sign of the dipole matrix elements given by Eq. (9.31).

nL_J	Experiments	Theory	Deviation
$6P_{1/2}$	**291.920(19)**[b]	317.9	8%
$6P_{3/2}$	**50.275(3)**[c]	48.3	4%
$7P_{1/2}$	**94.35(4)**[a]	99.9	6%
$7P_{3/2}$	**16.605(6)**[a]	15.4	7%
$8P_{1/2}$	**42.97(10)**[a]	45.0	5%
$8P_{3/2}$	**7.58(1)**[a]	7.0	7%
$9P_{1/2}$	–	24.2	–
$9P_{3/2}$	–	3.8	–
$10P_{1/2}$	–	14.5	–
$10P_{3/2}$	–	2.3	–
$11P_{1/2}$	–	9.4	–
$11P_{3/2}$	–	1.5	–

TABLE 9.2: Hyperfine coupling constants $A_{\text{hf}}(nL_J)$ in MHz. The experimental values are taken from (a) Arimondo et al. [24], (b) Rafac et al. [25], and (c) Tanner et al. [26]. The values marked in bold were used in the calculation of the third order tensor polarizability, where they contribute as diagonal matrix elements of H_{hf} to type B interactions. The theoretical values calculated from Schrödinger wavefunctions were used for states with $n > 8$. The theoretical values of the states $n = 6\ldots 8$ were not used in the calculation but are shown to illustrate the accuracy of our calculation. The hyperfine constants A_{nL_J} are related to the coupling constants $a(r)$ defined in the text by the relation $\langle a(r)\rangle = \frac{J(J+1)}{L(L+1)}A_{nL_J}$.

9.3 The third order polarizability of the free cesium atom   103

	$6P_{1/2}$	$7P_{1/2}$	$8P_{1/2}$	$9P_{1/2}$	$6P_{3/2}$	$7P_{3/2}$	$8P_{3/2}$	$9P_{3/2}$
$6P_{1/2}$	—	785.3	827.1	386.4	187.7	106.1	71.5	52.5
$7P_{1/2}$	—	—	422.2	309.4	105.2	85.0	57.3	42.1
$8P_{1/2}$	—	—	—	290.7	70.6	57.1	53.8	39.5
$9P_{1/2}$	—	—	—	—	51.8	41.8	39.4	41.3
$6P_{3/2}$	187.7	105.2	70.6	51.8	—	14.3	9.6	7.1
$7P_{3/2}$	106.1	85.0	57.1	41.8	—	—	7.8	5.7
$8P_{3/2}$	71.5	57.3	53.8	39.4	—	—	—	5.4
$9P_{3/2}$	52.5	42.1	39.5	41.3	—	—	—	—

TABLE 9.3: Off-diagonal hyperfine coupling constants $\langle a(r) \rangle$ (in MHz) representing interactions of type 2, 3 and 4. The numerical evaluation is based on Schrödinger wavefunctions as described in the text.

9.3.6 Terms with off-diagonal hf matrix elements

S-mixing off-diagonal hyperfine matrix elements have played an important role in the measurement of parity violation in the Cs atom and they were measured and calculated with a high accuracy. However, such matrix elements intervene only in type 1 interactions, and do thus not contribute to the present calculation, in which the relevant terms come from nP_J-$n'P_{J'}$ mixing (type 2,3, and 4) and $6S_{1/2} - mD_{3/2}$ mixing (type 5). We have evaluated all relevant diagonal and off-diagonal hyperfine matrix elements using Schrödinger wavefunctions and the expressions given above. A selection of numerical values of those matrix elements for the lowest-lying states is shown in Tables 9.3 and 9.4.

9.3.7 Contribution of continuum states

In principle the summation in Eq. (9.14) has to be carried out over continuum states as well as over bound states. It was shown in [14, 15] that the continuum states contribute significantly ($\sim 15\%$) to type 1 interactions, which themselves do not contribute to the tensor polarizability. In order to estimate the influence of the continuum on the tensor polarizability we calculated the (positive energy) continuum wavefunctions using our Schrödinger approach and used them to evaluate bound-continuum and continuum-continuum matrix elements of the electric dipole and hyperfine interactions.

For the numerical evaluation of transition matrix elements $\langle p, L'_{J'} | W | n L_J \rangle$ between bound $|n, L_J\rangle$ and continuum states $|p, L'_{J'}\rangle$ as well as between continuum states we use the expressions (9.32) and (9.28). We extended the perturbation sums (integrals) of all relevant interactions (types B, 2–5) over bound and continuum states.

The calculation poses no convergence problems, except for the type 5 interaction, which is the only one that involves dipole matrix elements between continuum states. Such matrix elements also occur, e.g., in the calculation of Bremsstrahlung transitions, and pose some technical difficulty. For their evaluation we have used the method of exterior complex scaling introduced in [27, 28] with a numerical implementation on a grid. The regular and outgoing waves in the complex plane were obtained through a step-by-step propagation using recursive relations [27].

As a result we find a total contribution from continuum states to the final value of $\alpha_2^{(3)}$ on the order of 0.2%.

9.3.8 Numerical evaluation of the third order tensor polarizability for the free Cs atom

We have done a numerical evaluation of all diagonal (type B) and off-diagonal (types 2,3,4, and 5) contributions to the tensor polarizability summing over bound states up to $n = 200$. For the $F = 4$ state we find

$$\alpha_2^{(3)}(F=4) = -3.72(25) \times 10^{-2} \text{ Hz/(kV/cm)}^2 , \quad (9.36)$$

in which the contribution from the quadrupole hyperfine interaction is 2×10^{-5} Hz/(kV/cm)^2. This result is shown as a dotted line in the left part of Fig. 9.4, together with previous theoretical and experimental results, described in the figure caption.

Table 9.5 shows the relative contributions of the different diagonal (type B) and off-diagonal (type 2–5) hyperfine interactions to the third order tensor polarizability. One notes that the diagonal (type B) contributions which were th only ones used in previous calculations overestimate the modulus of the third order tensor polarizability by approximately 50 %. Of all the off-diagonal contributions the interactions of type 2 and 5 (n-mixing of nP_J states with given J and $6S_{1/2}$-$nD_{3/2}$ mixing) give large contributions with opposite signs which cancel each other to a large extent.

We estimate the uncertainty of the final result to be ≈7%. This estimation is based on the precision (2–9%) with which our wavefunctions reproduce measured hyperfine constants (Table 9.2) and dipole matrix elements (Table 9.1), and considers that we have used (more precise) experimental values for the leading terms. As shown in the previous paragraph the relative contribution of the continuum states to the tensor polarizability is on the order of 10^{-3} and is thus negligible in the final result. Figure 9.1 shows the relative contributions from the different interactions in a cumulative way.

Our theoretical value for $\alpha_2^{(3)}(F=4)$ of the free Cs atom is in good agreement with all experimental values (Fig. 9.4).

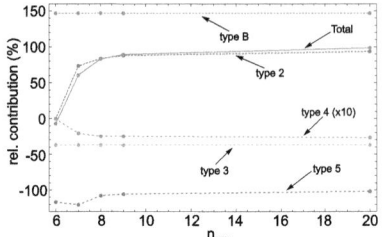

FIGURE 9.1: (Color online) Dependence of the different contributions to the perturbation sum (9.14) leading to the tensor polarizability on n_{max}, the maximum principal quantum number considered in the sum. Note that previous calculations took only diagonal (type B) hyperfine interactions into account.

ties $-4.07(20) \times 10^{-2}$ Hz/(kV/cm)^2 and $-4.36(28) \times 10^{-2}$ Hz/(kV/cm)^2 for $\alpha_2^{(3)}(F=4)$ are shown as points (d) and (e) in Fig. 9.4.

9.4 The third order polarizability of cesium in solid helium

9.4.1 Experiment

In connection with a proposed search for an electric dipole moment of Cs atoms in solid helium we were led to study the tensor Stark shifts in that unusual sample. The exceptionally long longitudinal and transverse spin relaxation times of Cs in the body-centered cubic (bcc) phase of ^{4}He crystals form the basis for high resolution magnetic resonance experiments. Recently we have measured the Stark shift of magnetic resonance lines in the ground state of Cs atoms implanted in bcc ^{4}He by two different techniques [2]. The experimental values of the corresponding tensor polarizabili-

n	$<mD_{3/2}|a(r)|6S_{1/2}>$
5	74.4
6	42.9
7	28.4
8	21.2
9	16.4

TABLE 9.4: Off-diagonal hyperfine coupling constants $\langle a(r) \rangle$ (in MHz) of type 5 interactions The numerical evaluation is based on Schrödinger wavefunctions as described in the text.

9.4.2 Wavefunction of Cs in solid ^{4}He

For calculating the tensor polarizability of Cs in solid helium one has to include the influence of the He matrix on the atomic energies and wavefunctions. Because of the Pauli repulsion Cs atoms form spherical bubbles, which are described by a spherically symmetric continuous He density distribution

$$\rho(R, R_0, \epsilon) = \begin{cases} 0 & R < R_0 \\ \rho_0 [1 - \{1 + \epsilon(R - R_0)\} e^{-\epsilon(R - R_0)}] & R \geq R_0 \end{cases}$$
(9.37)

where R_0 is the bubble radius. ϵ describes the steepness of the interface (density changes from zero to the bulk density) and ρ_0 is the bulk density which depends on the He temperature and pressure. In Eq. (9.37) we have assumed that solid helium is an incompressible fluid, an assumption well justified by the quantum nature of condensed ^{4}He. Our calculation of the Cs wavefunctions relies on an extension of the so-called bubble model [9], which was shown in the past to be well suited for describing energies, absorption/emission line shapes, hyperfine structure and lifetimes of alkalis in solid helium [29, 9].

The total interaction potential experienced by

9.4 The third order polarizability of cesium in solid helium

n type	6	7	8	9	10–200	total
B	146.8	≈0	≈0	≈0	≈0	146.8
2	–	73.4	10.2	4.3	6.4	94.3
3	-37.0	-0.1	≈0	≈0	≈0	-37.1
4	–	-2.1	-0.4	≈0	-0.2	-2.7
5	-116.7	-3.7	12.6	2.2	4.3	-101.3
total	-6.9	67.5	22.4	6.5	10.5	100.0

TABLE 9.5: Relative contributions (in %) to $\alpha_2^{(3)}(F=4)$ from diagonal, type B, Eq.(9.21) and off-diagonal, type 2–5, Eqs.(9.22) and (9.23), hyperfine interactions of excited states nL_J.

the Cs valence electron is given by

$$V_{Cs}^{bub}(r, R_0, \epsilon) = V_{Cs}(r) + \int d^3R\, \rho(R, R_0, \epsilon)$$
$$\times \left[V_{He}(\mathbf{r,R}) + V_{cross}(\mathbf{r,R}) + V_{cc}(\mathbf{R}) \right], \quad (9.38)$$

where $V_{Cs}(\mathbf{r})$ is the potential of the valence electron with the Cs$^+$ core introduced in Eq. (9.35). The ionic core and the He atoms are assumed to have fixed spatial positions (Born-Oppenheimer approximation). In Eq. (9.38) $V_{He}(\mathbf{r,R})$ and $V_{cc}(\mathbf{R})$ represent the interactions of the valence electron and the Cs$^+$ ion with a He atom. The potential $V_{cross}(\mathbf{r,R})$ describes the three-body interaction resulting from the simultaneous polarization of the He atom by the Cs core and the valence electron. $\mathbf{r}$ and $\mathbf{R}$ point from the core to the electron and to each He atom, respectively. Explicit forms for all the potentials as well as numerical parameter values are given in [9]. The potentials seen by the valence electron in the free Cs atom V_{Cs} and in the Cs atom trapped in solid He V_{Cs}^{bub} are shown, together with the energies of the lowest states in Fig. 9.3.

The energy needed to from a bubble consists of a pressure volume term, a surface energy (with surface tension parameter σ) and the kinetic energy E_{kin}, which arises from the localization of the He atoms at the bubble interface

$$E_{bub}(R_b, \epsilon) = \frac{4}{3}\pi R_b^3 p + 4\pi R_b^2 \sigma + E_{kin}, \quad (9.39)$$

where $R_b = f(R_0)$ is the center of gravity of the interface. The total energy of the bubble defect is thus $V_{tot}^{bub}(r, R_0, \epsilon) = V_{Cs}^{bub}(r, R_0, \epsilon) + E_{bub}(R_0, \epsilon)$, and the average bubble radius R_b is found by numerically minimizing the total energy with respect to the two parameters R_0 and ϵ. Using the known bubble parameters and the interaction potential one can then solve the radial Schrödinger equation as for the valence electron of the free Cs atom. In order to illustrate the effect of the He bubble we compare

in Fig. 9.2 the wavefunction of the $9P_{1/2}$ state of the free Cs atom with the corresponding wavefunction in the bubble. As expected, the wavefunction in the bubble is compressed due to the repulsive interaction with the surrounding helium atoms.

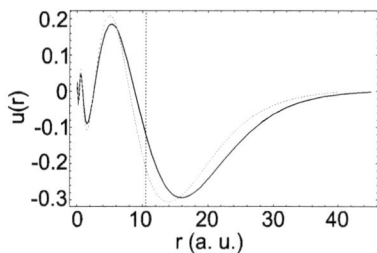

FIGURE 9.2: (Color online) Comparison of the $9P_{1/2}$ wavefunction of the free Cs atom (black solid line) with the same wavefunction of a Cs atom in a spherical helium bubble (red dotted line). The bubble parameters correspond to the equilibrium bubble shape of a ground state Cs atom ($R_0 = 10.2$, indicated by the vertical dotted line and $\epsilon = 2.45$ in atomic units).

Using the solutions (wavefunctions) one can then evaluate the matrix elements required for the numerical calculation of the third order polarizability, in analogy to the free atomic case.

9.4.3 Numerical evaluation of the third order tensor polarizability of Cs in solid He

The numerical evaluation of the tensor polarizability $\alpha_2^{(3)}$ for Cs in solid He was done in analogy to the case of the free atom, with wavefunctions and energy levels calculated using the bubble model introduced in Sect. 9.4.2. While in the case of the

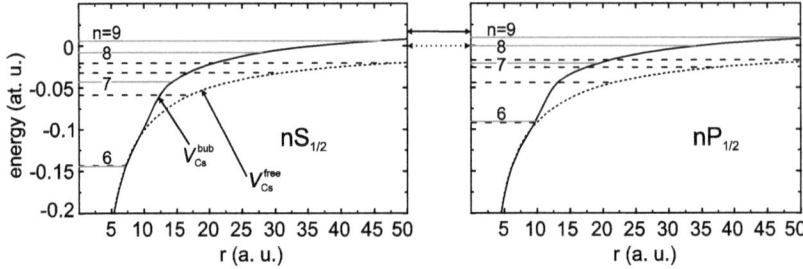

FIGURE 9.3: (Color online) Energies of the lowest $nS_{1/2}$ (left) and $nP_{1/2}$ (right) states of the free Cs atom (dashed lines) and of Cs in bcc solid ^{4}He (solid lines, $R_0 = 10.2$, $\epsilon = 2.45$ atomic units). The corresponding potentials for the valence electron of the free Cs atom [Eq. (9.35)] and the Cs atom in solid He [Eq. (9.38)] are shown as dotted and solid lines, respectively. The n=9 states are the highest bound states in the bubble. The bump in the potential V_{Cs}^{bub} is due to the repulsion by the bubble interface. The arrows between the graphs indicate the ionization limits.

	$n = 6, 7$	n=8	n=9	total
B	143.4	≈ 0	≈ 0	143.4
2	21.1	13.4	8.5	43.0
3	-26.8	≈ 0	≈ 0	-26.8
4	-0.2	-0.1	-0.1	-0.4
5	-65.4	4.1	2.1	-59.2
total	72.1	17.4	10.5	100.0

TABLE 9.6: Relative contributions (in %) to the tensor polarizability for Cs trapped in solid He. Note that Cs in solid He has no bound states with $n > 9$.

free Cs atom we have considered bound states up to the principal quantum number $n = 200$ we need only to consider states up to $9P$ in the He matrix, since higher lying states are not bound (see Fig. 9.3). Moreover, the energy levels of the higher lying states are strongly shifted to higher energies. For example the energies of the $8P_{1/2}$ and $9P_{1/2}$ state are displaced by 5580 cm^{-1} (22%) and 5830 cm^{-1} (21%), respectively, with respect to the free atom. Since the perturbation sum involves squared energy denominators these excited states give smaller contributions than in the free atomic case. In Table 9.6 we analyze the dependence of the relative contributions from the different diagrams on the number of states included in the perturbation sum which qualitatively reflects the same features as in the free atomic case. Our final theoretical value of the tensor polarizability of Cs in solid ^{4}He is

$$\alpha_2^{(3)}(F = 4) = -4.11 \times 10^{-2} \text{ Hz/(kV/cm)}^2. \quad (9.40)$$

On the right of Fig. 9.4 we compare this theoretical value with the experimental values of $\alpha_2^{(3)}$ of Cs atoms trapped in a body-centered cubic solid ^{4}He matrix [2] and find an excellent agreement.

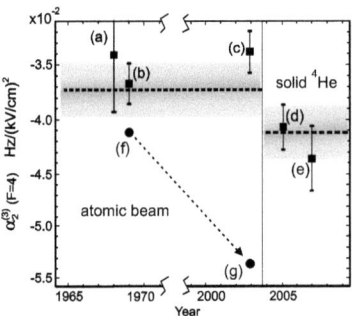

FIGURE 9.4: The Cs tensor polarizability $\alpha_2^{(3)}(F = 4)$. Atomic beam measurements of (a) Carrico et al. [7], (b) Gould et al. [6], (c) Ospelkaus et al. [8]. Points (d) and (e) represent recent measurements in solid helium [2]. The circles are the theoretical values (f) from Gould et al. [6] and (g) from Ulzega et al. [1]. The dashed lines represent the results of our new calculations for the free atom and for Cs in a solid helium matrix, together with their uncertainties (shaded bands).

9.5 Summary

We have presented calculational details of our semi-empirical evaluation of the cesium ground state tensor polarizability $\alpha_2^{(3)}$. This F- and M-dependent polarizability is forbidden in second order perturbation theory and arises only when considering the Stark effect and hyperfine interactions in a third order perturbation calculation. The tensor polarizability is suppressed by seven orders of magnitude with respect to the usual scalar polarizability of the atom.

We have evaluated $\alpha_2^{(3)}$ both for the free cesium atom and for Cs embedded in a cubic solid ^{4}He matrix using solutions of the Schrödinger equation with appropriate potentials for evaluating the relevant matrix elements in both cases. We have found that off-diagonal hyperfine matrix elements, which were not considered in previous treatments give substantial contributions (of different signs) to $\alpha_2^{(3)}$. As a result we obtain theoretical values for the tensor polarizabilities that are in excellent agreement with previous and recent measurements. The modulus of the experimental tensor polarizability of Cs in the He matrix is found to be $8 - 10\%$ larger than the one of the free Cs atom.

It thus seems that the 40 year old discrepancy between experimental and theoretical tensor polarizabilities has now found a satisfying final solution.

ACKNOWLEDGMENTS This work was supported by the grant number 200020-103864 of the Schweizerischer Nationalfonds.

References

[1] S. Ulzega, A. Hofer, P. Moroshkin, and A. Weis, Europhys. Lett. **76**, 1074 (2006).

[2] S. Ulzega, A. Hofer, P. Moroshkin, R. Müller-Siebert, D. Nettels, and A. Weis, Phys. Rev. A **75**, 042505 (2007).

[3] R. D. Haun and J. R. Zacharias, Phys. Rev. **107**, 107 (1957).

[4] E. Lipworth and P. G. H. Sandars, Phys. Rev. Lett. **13**, 716 (1964).

[5] P. G. H. Sandars, Proc. Phys. Soc. **92**, 857 (1967).

[6] H. Gould, E. Lipworth, and M. C. Weisskopf, Phys. Rev. **188**, 24 (1969).

[7] J. P. Carrico, A. Adler, M. R. Baker, S. Legowski, E. Lipworth, P. G. H. Sandars, T. S. Stein, and C. Weisskopf, Phys. Rev. **170**, 64 (1968).

[8] C. Ospelkaus, U. Rasbach, and A. Weis, Phys. Rev. A **67**, 011402(R) (2003).

[9] A. Hofer, P. Moroshkin, S. Ulzega, D. Nettels, R. Müller-Siebert, and A. Weis, Phys. Rev. A **76**, 022502 (2007).

[10] J. R. P. Angel and P. G. H. Sandars, Proc. R. Soc. London, A **305**, 125 (1968).

[11] J. M. Amini and H. Gould, Phys. Rev. Lett. **91**, 153001 (2003).

[12] A. Derevianko and S. G. Porsev, Phys. Rev. A **65**, 053403 (2002).

[13] H. L. Zhou and D. W. Norcross, Phys. Rev. A **40**, 5048 (1989).

[14] K. Beloy, U. I. Safronova, and A. Derevianko, Phys. Rev. Lett. **97**, 040801 (2006).

[15] E. J. Angstmann, V. A. Dzuba, and V. V. Flambaum, Phys. Rev. Lett. **97**, 040802 (2006).

[16] R. J. Rafac, C. E. Tanner, A. E. Livingston, K. W. Kukla, H. G. Berry, and C. A. Kurtz, Phys. Rev. A **50**, R1976 (1994).

[17] J. D. Feichtner, M. E. Hoover, and M. Mizushima, Phys. Rev. **137**, A702 (1965).

[18] A. A. Vasilyev, I. M. Savukov, M. S. Safronova, and H. G. Berry, Phys. Rev. A **66**, 020101(R) (2002).

[19] M. S. Safronova, W. R. Johnson, and A. Derevianko, Phys. Rev. A **60**, 4476 (1999).

[20] S. C. Bennett, J. L. Roberts, and C. E. Wieman, Phys. Rev. A **59**, R16 (1999).

[21] V. A. Dzuba, V. V. Flambaum, A. Y. Kraftmakher, and O. P. Sushkov, Phys. Lett. A **142**, 373 (1989).

[22] P. Gombas, *Pseudopotentiale* (Springer-Verlag Berlin-Gottingen-Heidelberg, 1956).

[23] D. W. Norcross, Phys. Rev. A **7**, 606 (1973).

[24] E. Arimondo, M. Inguscio, and P. Violino, Rev. Mod. Phys. **49**, 31 (1977).

[25] R. J. Rafac and C. E. Tanner, Phys. Rev. A **56**, 1027 (1997).

[26] C. E. Tanner and C. Wieman, Phys. Rev. A **38**, 1616 (1988).

[27] M. Chrysos and M. Fumeron, J. Phys. B **32**, 3117 (1999).

[28] M. Chrysos, J. Phys. B **33**, 2875 (2000).

[29] P. Moroshkin, A. Hofer, S. Ulzega, and A. Weis, Fiz. Nizk. Temp. **32**, 1297 (2006), (Low Temp. Phys. 32(11), 981 (2006)).

Summary and Outlook

In this thesis we have presented different studies of laser-excited Cs and Rb atoms implanted both in the cubic (bcc) and hexagonal phase (hcp) of polycrystalline ^{4}He matrices. In purely spectroscopic studies we have performed a detailed investigation of the He pressure dependence of the atomic excitation and emission spectra. Furthermore we have performed the first measurement of the $6P_{1/2}$ lifetime of Cs in bcc and hcp solid ^{4}He and studied its dependence on the pressure and crystal structure of the He matrix.

The experiments have yielded new insights into different decay channels of the atoms excited to lowest lying nP_J states. For example we have shown that the main decay channel of Rb atoms excited to the $5P_{1/2}$ or $5P_{3/2}$ is the formation of Rb*He$_6$ exciplexes in contrast to Cs, for which exciplexes are only formed in an efficient way following excitation to the $6P_{3/2}$ state. The $6P_{1/2}$ state of Cs was found to decay predominantly via emission of atomic fluorescence, and the exciplex formation channel slowly opens only in the hcp phase, probably due to the anisotropy of the local trapping site. From our lifetime measurements we could infer the pressure dependence of the exciplex formation probability.

Triggered by a several decades old discrepancy between experimental and theoretical values of the forbidden (tiny) tensor polarizabilities of alkali atoms, our group had remeasured the tensor polarizability in an atomic beam experiment several years ago. That measurement confirmed earlier measurements. In this work we have performed the first measurement of the tensor polarizability of Cs atoms in the cubic phase of solid ^{4}He, which was found to be approximately 10% larger than in the free Cs atom. In parallel Simone Ulzega had performed, in his Ph. D. thesis, a reanalysis of the theoretical calculations of that polarizability. After discovering and correcting for a sign error in the published treatment and after including off-diagonal matrix elements in the perturbation sum, a good agreement with experimental values was obtained. My contribution to that theoretical work was the setting up of a Schrödinger equation for the Cs valence electron, whose solutions, both for bound and continuum states, could then be used to evaluate the dipole and hyperfine matrix elements contributing to the perturbation sums. More recently I have included the effect of the He matrix in that calculation, which leads to a calculated tensor polarizability which is in good agreement with the measured ones. Based on our results a 40 year long history of discrepancy has now come to a final conclusion.

A major part of the thesis was devoted to the development of an extended bubble model, which allows the description of the trapping sites of alkali atoms in liquid and solid He matrices and of the influence of the He matrix on the atomic properties. In a complex mathematical treatment the potential experienced by the Cs valence electron under the influence of both the Cs core and the surrounding He bulk was derived. When inserted into the Schrödinger equation, that potential yields the perturbed energies and wavefunctions of the Cs atom. Those results were used to calculate the pressure shift of atomic absorption and emission lines, the alterations of the $6P_{1/2}$ lifetime in pressurized solid He, and the tensor polarizability, as already mentioned. In all cases the excellent agreement with our experimental results proves the power of this extended bubble model. Further more, previously neglected but necessary, refinements were included accounting for the cavity effect and for a new potential energy term resulting from elastic restoring forces from the He bulk.

In all of our experimental runs we had noted that during the melting of the He crystal at the end of the experiment a solid structure — which we named the *iceberg*, or the *blue finger* — corresponding to the doped region remains as a protrusion in the liquefied helium. In work initiated by P. Moroshkin, post-doc in our team, we have recently measured some of the physical properties of that structure, finding, e.g., that its coloration is due to nanometer-size alkali clusters and that its density lies somewhere between the densities of liquid and solid He. We tentatively explain the structure as being an aggregation of positively charged clusters which attract He to form a solid crust around them (so-called snowballs) and negative electron which repel helium and form bubbles.

The modelling of deformed bubbles, as they are known to occur in the hcp phase of solid He, will be one focus of our future theoretical activities. With such a model we hope to be able to explain, e.g., the experimentally observed discontinuities in the pressure dependence of absorption and emission lines as well as lifetimes at the bcc-hcp phase transition.

We have already performed first lifetime measurements of the different Cs exciplexes following D_2

excitation. However, because of severe cuts in funding from the Swiss National Science Foundation we could not purchase the equipment (photomultiplier for the near IR spectral range) necessary for extending those studies, which would yield more information on the exciplex formation processes and the involved time scales. In another line of research we plan to study the Hanle effect of excited Cs atoms. Such level-crossing signals, together with the knowledge of the excited state lifetime, yield information about the g-factor of excited alkali states in solid He. Equipment was already purchased or manufactured for those measurements.

Furthermore we will extend our investigation of the iceberg structure, by performing measurements of the electric current induced by a static electric field. From that study we hope to glean information on the density of charged particles in that structure, a further step for giving more weight to its tentative interpretation in terms of an ionic quantum crystal.

Die VDM Verlagsservicegesellschaft sucht für wissenschaftliche Verlage abgeschlossene und herausragende

Dissertationen, Habilitationen, Diplomarbeiten, Master Theses, Magisterarbeiten usw.

für die kostenlose Publikation als Fachbuch.

Sie verfügen über eine Arbeit, die hohen inhaltlichen und formalen Ansprüchen genügt, und haben Interesse an einer honorarvergüteten Publikation?

Dann senden Sie bitte erste Informationen über sich und Ihre Arbeit per Email an *info@vdm-vsg.de*.

Sie erhalten kurzfristig unser Feedback!

VDM Verlagsservicegesellschaft mbH
Dudweiler Landstr. 99   Telefon +49 681 3720 174
D - 66123 Saarbrücken   Fax   +49 681 3720 1749
www.vdm-vsg.de

Die VDM Verlagsservicegesellschaft mbH vertritt

Printed by Books on Demand GmbH, Norderstedt / Germany